Lecture Notes in Control and Information Sciences

Edited by A. V. Balakrishnan and M. Thoma

Vol. 22: Optimization Techniques
Proceedings of the 9th IFIP Conference on
Optimization Techniques,
Warsaw, September 4–8, 1979
Part 1
Edited by K. Iracki, K. Malanowski, S. Walukiewicz
XVI, 569 pages. 1980

Vol. 23: Optimization Techniques
Proceedings of the 9th IFIP Conference on
Optimization Techniques,
Warsaw, September 4-8, 1979
Part 2
Edited by K. Iracki, K. Malanowski, S. Walukiewicz
XV, 621 pages. 1980

Vol. 24: Methods and Applications
in Adaptive Control
Proceedings of an International Symposium
Bochum, 1980
Edited by H. Unbehauen
VI, 309 pages. 1980

Vol. 25: Stochastic Differential Systems –
Filtering and Control
Proceedings of the IFIP-WG7/1 Working Conference
Vilnius, Lithuania, USSR, Aug. 28 – Sept. 2, 1978
Edited by B. Grigelionis
X, 362 pages. 1980

Vol. 26: D. L. Iglehart, G. S. Shedler
Regenerative Simulation of Response
Times in Networks of Queues
XII, 204 pages. 1980

Vol. 27: D. H. Jacobson, D. H. Martin, M. Pachter, T. Geveci
Extensions of Linear-Quadratic Control Theory
XI, 288 pages. 1980

Vol. 28: Analysis and Optimization of Systems
Proceedings of the Fourth International
Conference on Analysis and Optimization of Systems
Versailles, December 16–19, 1980
Edited by A. Bensoussan and J. L. Lions
XIV, 999 pages. 1980

Vol. 29: M. Vidyasagar,
Input-Output Analysis of Large-Scale
Interconnected Systems –
Decomposition, Well-Posedness and Stability
VI, 221 pages. 1981

Vol. 30: Optimization and Optimal Control
Proceedings of a Conference Held at
Oberwolfach, March 16–22, 1980
Edited by A. Auslender, W. Oettli, and J. Stoer
VIII, 254 pages. 1981

Vol. 31: Berc Rustem
Projection Methods in Constrained
Optimisation and Applications
to Optimal Policy Decisions
XV, 315 pages. 1981

Vol. 32: Tsuyoshi Matsuo,
Realization Theory of
Continuous-Time Dynamical Systems
VI, 329 pages, 1981

Vol. 33: Peter Dransfield
Hydraulic Control Systems –
Design and Analysis of Their Dynamics
VII, 227 pages, 1981

Vol. 34: H.W. Knobloch
Higher Order Necessary Conditions
in Optimal Control Theory
V, 173 pages, 1981

Vol. 35: Global Modelling
Proceedings of the IFIP-WG 7/1 Working
Conference Dubrovnik, Yugoslavia,
Sept. 1–5, 1980
Edited by S. Krčevinac
VIII, 232 pages, 1981

Vol. 36: Stochastic Differential Systems
Proceedings of the 3rd IFIP-WG 7/1
Working Conference
Visegrád, Hungary, Sept. 15–20, 1980
Edited by M. Arató, D. Vermes, A.V. Balakrishnan
VI, 238 pages, 1981

Vol. 37: Rüdiger Schmidt
Advances in Nonlinear
Parameter Optimization
VI, 159 pages, 1982

Vol. 38: System Modeling and Optimization
Proceedings of the 10th IFIP Conference
New York City, USA, Aug. 31 – Sept. 4, 1981
Edited by R.F. Drenick and F. Kozin
XI, 894 pages. 1982

Vol. 39: Feedback Control of
Linear and Nonlinear Systems
Proceedings of the Joint Workshop
on Feedback and Synthesis of
Linear and Nonlinear Systems
Bielefeld/Rom
XIII, 284 pages. 1982

Vol. 40: Y.S. Hung, A.G.J. MacFarlane
Multivariable Feedback:
A Quasi-Classical Approach
X, 182 pages. 1982

Vol. 41: M. Gössel
Nonlinear Time-Discrete Systems –
A General Approach by
Nonlinear Superposition
VIII, 112 pages. 1982

Vol. 42: Advances in Filtering and
Optimal Stochastic Control
Proceedings of the IFIP-WG 7/1
Working Conference
Cocoyoc, Mexico, February 1–6, 1982
VIII, 391 pages. 1982

For information about Vols. 1–21 please contact your bookseller or Springer-Verlag.

Lecture Notes in Control and Information Sciences

Edited by M. Thoma and A. Wyner

82

Analysis and Algorithms of Optimization Problems

Edited by
K. Malanowski, K. Mizukami

Springer-Verlag Berlin Heidelberg GmbH

Editors
K. Malanowski
Systems Research Institute of the Polish Academy of Sciences
ul. Newelska 6
01-447 Warszawa
Poland

K. Mizukami
Hiroshima University
Faculty of Integrated Arts and Sciences
Higashisenda-machi
Hiroshima 730
Japan

ISBN 978-3-540-16660-3 ISBN 978-3-540-39844-8 (eBook)
DOI 10.1007/978-3-540-39844-8

Originally published by Springer-Verlag Berlin Heidelberg New York in 1986.

2161/3020-543210

PREFACE

In 1981 a Japanese-Polish cooperation in the subject "Numerical Methods of Optimization and Game Theory" was established. It was sponsored and supported respectively by the Japanese Society for Promotion of Science and by the Polish Academy of Sciences.

This cooperation involved scientists from Hiroshima and Osaka Universities on the Japanese side and the Systems Research Institute of the Polish Academy of Sciences on the Polish side.

The cooperation resulted, among others, in a number of joint papers and a book "Constructive Aspects of Optimization", Państwowe Wydawnictwo Naukowe (Polish Scientific Publishers), Warszawa-Łódź, 1985.

Formally the cooperation ended in 1983 but it has been continued on informal basis and the present book is its result.

Among the authors contributing to the book Professor Koichi Mizukami and Mr. Naofumi Iwata are with the Department of the Information and Behavioral Sciences of the Hiroshima University, Professor Yoshiyuki Sakawa and Dr. Yuji Shindo are with the Department of Control Engineering of the Osaka University, Professor Constantin Vârsan is with the Department of Mathematics, National Institute for Scientific and Technical Creation, Bucharest, Romania, while all remaining contributors are from the Systems Research Institute of the Polish Academy of Sciences.

K. Malanowski K. Mizukami

INTRODUCTION

This book consists of ten independent and self-contained chapters written by different authors. Most of the presented results belong to the authors themselves. They are either published here for the first time or having been partially published elsewhere they are presented here in a different form. The proofs of the results are sometime omitted, but the detailed references are provided.

The main part of the presented material (Chapter 1 through 6) is devoted to modelling and optimization of distributed parameter systems. Thus, among them the three first chapters concern theoretical aspects of optimization, while in the next three some numerical problems are presented.

The material presented in Chapter 7 is very close to that in the previous part, namely there is proposed an iterative algorithm of solving some optimal control problems for systems described by ordinary differential equations.

Chapters 8 and 9 are devoted to some game-theoretical problems. Finally, Chapter 10 concerns calculation of the so called surrogate constraints in mathematical programming problems.

A short outline of the results presented in all chapters is given below.

Chapters 1 and 2 concern sensitivity analysis of solutions to optimal control problems for distributed parameter systems.

More precisely, the dependence of solutions on a parameter, which enters the data of convex optimal control problems subject to inequality-type constraints is investigated. It is shown that in the considered cases the solutions to the optimization problems, as functions of the parameter, are directionally (conically) differentiable and the respective right-derivatives can be found effectively as the solutions to auxiliary quadratic optimal control problems. Sufficient conditions under which these functions are Gâteaux differentiable are obtained.

In this analysis the main difficulty is created by the presence of inequality type constraints. In Chapters 1 and 2 two different methods of coping with this difficulty are presented.

In Chapter 1 the results of the directional differentiability of the mapping of projection onto a closed and convex set in a Hilbert space are exploited to obtain differentiability of solutions for some state and control constrained optimal control problems with linear constraints.

In Chapter 2 the Lagrange formalism for optimal control problems subject to convex pointwise constraints is used. This approach allows to analyse the differential properties of both the solutions and the associated Lagrange multipliers, it requires however, Lipschitz continuity results for both these functions. For the considered problem Lipschitz continuity is proved.

Chapter 3 concerns problems of parametric optimal control for linear evolution equations and some free boundary problems. In this class of problems control is executed through coefficients of the involved elliptic operators.

Since these optimal control problems may not have solutions a concept of generalized solution is introduced. This cencept is based on the notion of the so called G-convergence of operators.

The results concerning G-convergence of the second order elliptic operators are presented with emphasis on isotropic operators. They are used to define generalized parametric optimal control problems for parabolic equations and variational inequalities. These generalized problems have solutions and for them necessary conditions of optimality are presented.

One of the most important areas, where parametric optimization problems occur in practice is optimal design of mechanical structures.

Chapters 4 and 5 are devoted to numerical methods for solving some optimal design problems.

Chapter 4 deals with optimal design of a plate with respect the fundamental frequency of its free vibrations. The volume of the plate is fixed, while its thickness is subject to optimization. The optimization problem consists in maximizing the smallest eigenvalue of the fourth order elliptic eigenvalue problem describing free vibrations of the plate.

To approximate this problem the finite element method is employed. The convergence of approximation is proved.

The discretized problem is nonsmooth in the case where the smallest eigenvalues are multiple, therefore to solve it a method of nondifferentiable optimization is used. Numerical examples are presented.

In Chapter 5 an optimal shape design problem for two-dimensional elastic body, subject to external forces is investigated. Like in the previous papers by the author the approach used is based on direct minimization of the performance index with respect to some shape parameters treated as decision variables. However, in contrast to the previous papers, perforated domains are considered. A method of homogenization is applied, allowing to approximate the orginal problem with a reasonable accuracy.

A two level design of the shape is proposed and it is illustrated by numerical examples.

Chapter 6 concerns numerical methods for solving multiphase problems of Stefan type in several space variables.

The method exploits the fixed domain formulation of the problems in the form of variational inequalities of the parabolic or mixed elliptic-parabolic type. For this formulation stable approximation schemes are constructed using finite elements in space and finite differences in time variables. The schemes provide a simple time-stepping algorithm. Presented results of numerical experiments indicate the efficiency of the proposed algorithm both for parabolic and for degenerate elliptic-parabolic Stefan problems.

Chapter 7 presents a modification and simplification of an efficient numerical algorithm of solving optimal control problems for systems described by nonlinear ordinary differential equations where the cost functional depends both on the terminal state and the whole trajectory. The original algorithm was developed by the author and Y. Shindo.

Chapter 8 is devoted to effective construction of a quasi-optimal feedback solution for linear differential games, without the necessity of solving the partial differential equation associated with the optimal strategy. The analysis is performed from the point of view of the first player. Both deterministic and stochastic cases are considered.

In the deterministic case construction of a quasi-optimal feedback requires the knowledge of the strategy used by the second player. In this case a numerical example is provided.

In the stochastic case it is allowed that the second player uses nonanticipating processes as the admissible strategies, and it is shown that the analitical form of the quasi-optimal feedback for the first player is independent of the strategy used by the second player.

In Chapter 9 the feedback Nash equilibrium strategies are considered for continuous-time, deterministic two-person differential game with a nonlinear state equation and quadratic cost functionals. The nonlinearity of the state equation appears as a regular perturbation.

The optimal feedback strategy is obtained in the form of a series. The elements of the series can be calculated by solving a matrix Riccati equation and a sequence of quasi-linear partial differential equations.

Several theorems concerning the asymptotic properties of the approximations of the Nash equilibrium strategies are included.

Chapter 10 deals with calculating surrogate constraints mainly for integer programming problems. It is well known that the surrogate dual problems can offer effective bounds on the primal optimal values. Knowledge of these bounds is of a great importance to any branch-and-bound algorithm. However, solving of the dual problems is rather difficult since it requires maximizing of a quasi-concave, often discontinuous, function.

A certain method for calculating surrogate constraints is analysed theoretically and numerically.

The proposed algorithm is based on the concept of the quasi-subgradient generalizing the notion of the subgradient for quasi-concave functions.

The convergence of the algorithm is proved and some numerical results are presented.

CONTENTS

Chapter 1

DIFFERENTIAL STABILITY OF PROJECTION IN HILBERT SPACE ONTO CONVEX SET. APPLICATIONS TO SENSITIVITY ANALYSIS OF OPTIMAL CONTROL PROBLEMS

Jan Sokołowski

1. Introduction

The paper is concerned with the differential stability of solutions of variational inequalities with respect to the parameter. The first part of the paper is devoted to the differential stability of the projection in Hilbert space onto a closed and convex subset. We exploit the notion of the conical differentiability of the projection mapping. Using the results on conical differentiability of the projection we derive the form of the sensitivity coefficient of an optimal control with respect to the parameter for the constrained optimal control problems for distributed parameter systems.

We start with the following examples.

Example 1.1

Let us consider an elementary example of the projection mapping $P_K(.)$ in R onto the set $K=[0,+\infty)$.
In the case we have

$$\forall x \in R : P_K(x) = x^+ = \begin{cases} x , & x \geq 0 \\ 0 , & x < 0 \end{cases} \tag{1.1}$$

It is easy to see that the mapping $x \to x^+$ is differentiable everywhere except at $x=0$. At the point $y=0$ we have for $h=\pm 1$ and for $\varepsilon > 0$:

$$(y + \varepsilon h)^+ = y^+ + \varepsilon h^+ \tag{1.2}$$

therefore for $\varepsilon > 0$

$$[(y + \varepsilon h)^+ - y^+]/\varepsilon = h^+ = \lim_{\varepsilon \downarrow 0}[(y + \varepsilon h)^+ - y^+]/\varepsilon \tag{1.3}$$

hence at $y=0$ we have

$$P_K(y + \varepsilon h) = P_K(y) + \varepsilon Q(h) + o(\varepsilon) \tag{1.4}$$

where the mapping $Q(.) : R \to R$ is defined by $Q(h)=h^+$, $\forall h \in R$.

In the notation of the paper the mapping $Q(.)$ is called the conical differential of the projection $P_K(.)$ at $y=0$.

Let us recall how the projection mapping $P_K(.)$ is related to the variational inequalities. Since for a given $x \in R$ we have

$$(x^+ - x)^2 \leq (v - x)^2, \quad \forall v \in K \tag{1.5}$$

then by a standard argument it follows that the element $x^+=P_K(x)$ is given by unique solution of the following variational inequality:

$$x^+ \in K : (x^+ - x)(v - x^+) \geq 0, \quad \forall v \in K \tag{1.6}$$

$\blacksquare$

In this paper we will use the results on differential stability of the projection in a Hilbert space for the local sensitivity analysis of optimal controls to constrained, convex optimal control problems depending on the parameter. Let us show how the differential stability of optimal controls is related to the differential stability of solutions of variational inequalities. We present a simple example of the optimal control problem for ordinary differential equation.

Example 1.2

We denote by $L^2(0,T)$ the space of square integrable functions on $(0,T)$, $T > 0$. $L^2(0,T)$ is Hilbert space with the scalar product

$$(y,z)_{L^2(0,T)} = \int_0^T y(t)\, z(t)dt, \quad \forall y,z \in L^2(0,T) \tag{1.7}$$

We denote by $H^1(0,T)$ Sobolev space:

$$H^1(0,T) = \{\phi \in L^2(0,T) \mid \frac{d\phi}{dt} \in L^2(0,T)\} \tag{1.8}$$

Space $H^1(0,T)$ is Hilbert space with the scalar product

$$(y,z)_{H^1(0,T)} = \int_0^T \{y(t)z(t) + \frac{dy}{dt}(t)\frac{dz}{dt}(t)\}dt \tag{1.9}$$

In order to define an optimal control problem we introduce the state equation, the cost functional and the set of admissible states of the form:

<u>state equation</u>:

$$\frac{dy}{dt}(t) = y(t) + u(t), \quad t \in (0,T)$$
$$y(0) = 0 \tag{1.10}$$

$u(.) \in L^2(0,T)$ denotes control

$y(.) \in H^1(0,T)$ denotes state

<u>cost functional</u>:

$$J(u) = \frac{1}{2} \int_0^T (y(t) - y_d(t))^2 dt + \frac{\alpha}{2} \int_0^T (u(t))^2 dt \tag{1.11}$$

$\alpha > 0$, where $y_d(.) \in L^2(0,T)$ is given element

<u>set of admissible states</u>:

$$Y_{ad} = \{y(.) \in H^1(0,T) \mid y(0)=0, \ a \le y(T) \le b\} \tag{1.12}$$

where $a,b \in R$ are given constants.

We denote by $u_o \in L^2(0,T)$ an optimal control which minimizes the cost functional (1.11) subject to state equation (1.10) and state constraints (1.12); we denote by $y_o(.) \in H^1(0,T)$ the optimal state. Let us consider the differential stability of the mapping

$$L^2(0,T) \ni y_d \longrightarrow u_o \in L^2(0,T) \tag{1.13}$$

Let $h(.) \in L^2(0,T)$ be a given element, denote by $u_\varepsilon \in L^2(0,T)$, $\varepsilon \in [0,\delta)$, $\delta > 0$, an optimal control which minimizes the cost functional

$$J_\varepsilon(u) = \frac{1}{2} \int_0^T (y(t) - y_d(t) - \varepsilon h(t))^2 dt + \frac{\alpha}{2} \int_0^T (u(t))^2 dt$$

subject to state equation (1.10) and state constraints (1.12). Denote by $y_\varepsilon \in H^1(0,T)$ the optimal state given by a unique solution of the state equation:

$$\frac{dt_\varepsilon}{dt}(t) = y_\varepsilon(t) + u_\varepsilon(t), \quad t \in (0,T) \tag{1.14}$$

$$y_\varepsilon(0) = 0 \tag{1.15}$$

It can be verified that the optimal state is given by a unique solution of the following variational inequality:

find an element $y_\varepsilon \in K$ such that

$$a(y_\varepsilon,\phi-y_\varepsilon) \geq \int_0^T y_d(t)(\phi(t)-y_\varepsilon(t))dt \qquad (1.16)$$
$$\forall\phi \in K$$

where $K \overset{def}{=} Y_{ad}$ and the bilinear form $a(.,.) : H^1(0,T) \times H^1(0,T) \to R$ is defined as follows

$$a(y,z) \overset{def}{=} \int_0^T \{\alpha\dot{y}(t)\dot{z}(t)-\alpha y(t)\dot{z}(t)-\alpha\dot{y}(t)z(t)+(1+\alpha)y(t)z(t)\}dt\ ,$$
$$\forall y,z \in H^1(0,T) \qquad (1.17)$$

here we denote $\dot{y}=dy/dt$.

We can apply the results on differential stability of the metric projection in Hilbert space presented in the paper to the variational inequality (1.16). It follows that for $\varepsilon > 0$, ε small enough

$$y_\varepsilon = y_o + \varepsilon z + o(\varepsilon) \quad \text{in} \quad H^1(0,T) \qquad (1.18)$$

where $\|o(\varepsilon)\|_{H^1(0,T)}/\varepsilon \to 0$ with $\varepsilon \downarrow 0$. The element $z \in H^1(0,T)$ is given by a unique solution of the following variational inequality:

find an element $z \in S$ such that

$$a(z,\phi-z) \geq \int_0^T h(t)(\phi(t)-z(t))dt\ , \quad \forall\phi \in S \qquad (1.19)$$

where the cone S is given by

$$S = \{\phi \in H^1(0,T) \mid \phi(0) = 0, \qquad (1.20)$$
$$\phi(T) \geq 0 \quad \text{if} \quad y_o(T) = a,$$
$$\phi(T) \leq 0 \quad \text{if} \quad y_o(T) = b,$$
$$a(y_o,\phi) = \int_0^T y_d(t)\phi(t)dt\}$$

From (1.18) and (1.14) it follows that for $\varepsilon > 0$, ε small enough:

$$u_\varepsilon = u_o + \varepsilon q + o(\varepsilon) \quad \text{in} \quad L^2(0,T) \qquad (1.21)$$

where $\|o(\varepsilon)\|_{L^2(0,T)}/\varepsilon \to 0$ with $\varepsilon \downarrow 0$.

It can be verified that the element $q \in L^2(0,T)$ is given by a unique solution of the following optimal control problem:

find an element $q \in L^2(0,T)$ which minimizes the cost functional

$$I(u) = \frac{1}{2} \int_0^T (z(t)-h(t))^2 dt + \frac{\alpha}{2} \int_0^T (u(t))^2 dt \tag{1.22}$$

subject to state equation (1.10) and state constraints:

$$z(T) \leq 0 \quad \text{if} \quad y_o(T) = b \tag{1.23}$$

$$z(T) \geq 0 \quad \text{if} \quad y_o(T) = a \tag{1.24}$$

$$a(y_o, z) = \int_0^T y_d(t) z(t) dt \tag{1.25}$$

■

The element q in (1.21) is called the sensitivity coefficient for the optimal control u_o. The Example 1.2 shows that the sensitivity coefficient for an optimal control can be obtained in the form of an optimal solution of the auxiliary optimal control problem.

The differential stability of solutions of variational inequalities with respect to the perturbations of the right-hand side has been studied by Mignot [19] and Haraux [7]. In [19] the notion of a polyhedric convex subset of Hilbert space is introduced and the form of the so-called conical differential of the projection onto such a subset is derived.

Several results on differential stability of metric projection in Hilbert space onto convex set are given by Holmes [8] and by Fitzpatrick and Phelps [5], we refer the reader also to [37] for the related results.

The differential stability of solutions to constrained mathematical programming problems is investigated e.g. in [4, 9, 15]. The results presented in [15] has been used in [16, 17] in order to derive the form of the right-derivatives of solutions to convex, constrained optimal control problems for systems described by ordinary differential equations. Sensitivity analysis of the constrained optimal control problems for partial differential equations is considered in [18] using the similar method as in [16, 17].

In this paper the method proposed by the author [24, 25, 26, 29] based on the conical differentiability of the projection is used in order to derive the form of the right-derivative of an optimal control for optimal control problems for distributed parameter systems with respect to the parameter. In this chapter the right - derivative of an optimal control is called the sensitivity coefficient of an optimal control with respect to the parameter.

The main result which is used in our method of the sensitivity analysis [24, 26] is the following: the sensitivity coefficient of an optimal solution with respect to the parameter can be derived in the form of an optimal solution of an auxiliary constrained optimization problem.

For further results on differential stability of solutions to variational inequalities as well as on the sensitivity analysis of the optimal control problems we refer the reader to [22, 23, 27, 28]. In [30-36] the applications to the shape sensitivity analysis of free boundary problems are given. The related results on the shape sensitivity analysis of optimal control problems are presented in [25, 26, 29].

The outline of this chapter is following. In Section 2 the projection mapping in Hilbert space onto convex, closed subset is considered. The notion of the conical differentiability of the mapping is introduced. In Section 3 an abstract result on conical differentiability of the projection mapping is presented. Section 4 is concerned with the differential stability of solutions to an abstract, constrained optimization problem. An example of constrained optimal control problem is provided.

Finally in Section 5 the results on differential stability of optimal controls for two examples are presented.

In the paper the standard notation is used [11]. The related results concerning variational inequalities and optimal control problems can be found in [3, 6, 10, 12, 13, 14, 21].

We use the following notation [11].

Let $\Omega \subset R^n$ be a given domain with the smooth boundary $\Gamma = \partial\Omega$. We denote by $L^2(\Omega)$ the space of square integrable functions on Ω. $L^2(\Omega)$ is Hilbert space with scalar product of the form:

$$(y,z)_{L^2(\Omega)} = \int_\Omega y(x)z(x)dx \ , \quad \forall y,z \in L^2(\Omega) \tag{1.26}$$

We denote by $H^1(\Omega)$, $H^2(\Omega)$ Sobolev spaces:

$$H^1(\Omega) = \{\phi \in L^2(\Omega) \mid \frac{\partial\phi}{\partial x_i} \in L^2(\Omega), \quad i=1,\dots,n\} \tag{1.27}$$

$$H^2(\Omega) = \{\phi \in L^2(\Omega) \mid \frac{\partial\phi}{\partial x_i}, \frac{\partial^2\phi}{\partial x_i \partial x_j} \in L^2(\Omega), \quad i,j=1,\dots,n\} \tag{1.28}$$

Spaces $H^1(\Omega)$, $H^2(\Omega)$ are Hilbert spaces [11] with the scalar products:

$$(y,z)_{H^1(\Omega)} = \int_\Omega \{y(x)z(x) + \nabla y(x).\nabla z(x)\}dx$$

here we denote $\nabla y(x) = \mathrm{col}\,(\frac{\partial y}{\partial x_1},\dots,\frac{\partial y}{\partial x_n})$

$$(y,z)_{H^2(\Omega)} = \int_\Omega \{y(x)z(x)+\nabla y(x).\nabla z(x)+\Delta y(x).\Delta z(x)\}dx$$

where $\Delta y = \mathrm{div}(\nabla y) = \sum_{i=1}^{n} \frac{\partial^2 y}{\partial x_i^2}$.

Sobolev space $H_0^1(\Omega)$ is defined as follows [11]:

$$H_0^1(\Omega) = \{\phi \in H^1(\Omega) \mid \phi(x) = 0 \quad \text{on} \quad \partial\Omega\} \tag{1.29}$$

It is Hilbert space with the scalar product:

$$(y,z)_{H_0^1(\Omega)} = \int_\Omega \nabla y(x).\nabla z(x)dx \tag{1.30}$$

2. Projection mapping in Hilbert space

Let H be a separable Hilbert space, $K \subset H$ a convex and closed subset. Let there be given a bilinear form

$$a(.,.) : H \times H \to R \tag{2.1}$$

which is coercive and continuous i.e.,

$$a(v,v) \geq \alpha \|v\|_H^2, \quad \alpha > 0, \quad \forall v \in H \tag{2.2}$$

$$|a(v,z)| \leq M \|v\|_H \|z\|_H, \quad \forall v,z \in H \tag{2.3}$$

Let H' denotes the dual space of H and let $f \in H'$ be a given element. We denote by $y=P(f)$ a unique solution of the variational inequality:

$$\begin{aligned} &y = P(f) \in K \\ &a(y,v-y) \geq \langle f,v-y \rangle, \quad \forall v \in K \end{aligned} \tag{2.4}$$

where $\langle .,. \rangle$ is the a duality pairing between H' and H.

Remark 2.1:

If the bilinear form $a(.,.)$ is symmetric i.e., $a(v,z)=a(z,v)$, $\forall v,z \in H$ then

$$y = P(f) = \arg\min\{\frac{1}{2} a(v,v) - \langle f,v \rangle \mid v \in K\} \tag{2.5}$$

■

It can be verified that the mapping

$$H' \ni f \longrightarrow P(f) \in H \tag{2.6}$$

is Lipschitz continuous:

$$\|P(f_1) - P(f_2)\|_H \leq \frac{M}{\alpha} \|f_1 - f_2\|_{H'}, \quad \forall f_1, f_2 \in H' \tag{2.7}$$

therefore by a generalization of the Rademacher theorem [19] it follows that there exists a dense subset $\Xi \subset H'$ such that for $f \in \Xi$ we have

$$\forall h \in H' : P(f + \varepsilon h) = P(f) + \varepsilon P'(h) + r(\varepsilon) \quad \text{in} \quad H \tag{2.8}$$

where $r(\varepsilon)/\varepsilon \to 0$ strongly in H with $\varepsilon \to 0$.
The mapping $P'(.)=P'(f;.) : H' \to H$ is linear and continuous. In the sequal we will use the concept of the so-called conical differentiability of the projection operator.

Definition 2.1

The mapping (2.6) is conically differentiable at $f \in H'$ if there exists a continuous mapping

$$Q(.) : H' \longrightarrow H \tag{2.9}$$

such that for $\varepsilon > 0$, ε small enough

$$\forall h \in H' : P(f + \varepsilon h) = P(f) + \varepsilon Q(h) + o(\varepsilon) \quad \text{in} \quad H \tag{2.10}$$

where $\|o(\varepsilon)\|_H/\varepsilon \to 0$ with $\varepsilon \downarrow 0$ uniformly on compact subsets of H'.
In order to derive the form of the mapping (2.9) we need the following notation.

For a given element $y \in K$ we denote by $C_K(y)$ the tangent cone

$$C_K(y) = \{\phi \in H \mid \exists\, \varepsilon > 0 \text{ such that } y + \varepsilon\phi \in K\} \tag{2.11}$$

In general the cone (2.11) is not closed, we denote by $\overline{C_K(y)}$ its closure in H.

For a given element $f \in H'$ we denote by $T_K(f) \subset H$ a hyperplane of the form:

$$T_K(f) = \{\phi \in H \mid a(P_K(f),\phi) = \langle f,\phi \rangle\} \quad (2.12)$$

Finally we denote by $S_K(f) \subset H$ a convex closed cone of the form:

$$S_K(f) = \overline{C_K(P_K(f))} \cap T_K(f)\ , \quad \forall f \in H' \quad (2.13)$$

We present several results concerning the differential stability of the projection $P_K(.)$.

Lemma 2.1. [7]

Assume that bilinear form $a(.,.)$ is symmetric. Denote

$$\gamma(\varepsilon)=(P_K(f+\varepsilon h) - P_K(f))/\varepsilon \quad (2.14)$$

where $f,h \in H'$ are given elements, $\varepsilon > 0$.
Every weak limit γ in H of the sequence $\{\gamma(\varepsilon)\}$ for $\varepsilon \downarrow 0$ verifies the following condition

$$\gamma \in S_K(f) \quad (2.15)$$

Lemma 2.2

For a given element $f \in H'$ denote by $F \in H$ a unique solution of the following variational equation:

$$a(F,\phi) = \langle f,\phi \rangle\ , \quad \forall \phi \in H \quad (2.16)$$

If $F \in K$ then the mapping (2.6) is conical differentiable and we have

$$Q(h) = P_S(h)\ , \quad \forall h \in H' \quad (2.17)$$

where $S = \overline{C_K(f)}$.
The proof of Lemma 2.2 follows from the results presented in [37].

Definition 2.2

The set $K \subset H$ is called polyhedric if for any $f \in H'$ it follows that

$$S_K(f) = \overline{C_K(P_K(f)) \cap T_K(f)} \quad (2.18)$$

Example 2.1

We set

$$H = R^m$$

$$a(y,z) = \sum_{i=1}^{m} y_i z_i \ , \quad \forall y,z \in R^m \ , \quad y = \mathrm{col}(y_1,\dots,y_m)$$

It can be verified that the set $K \subset R^m$ of the form:

$$K = \{y \in R^m \mid \langle a^i, y \rangle_{R^m} + b^i \leq 0 \ , \quad i=1,\dots,N\} \tag{2.19}$$

is polyhedric. Here $a^i \in R^m$, $b^i \in R$ are given elements for $i=1,\dots,N$.

Example 2.2

The cone

$$K = \{y \in L^2(\Omega) \mid y(x) \geq 0 \quad \text{a.e. in} \quad \Omega\} \tag{2.20}$$

is polyhedric [27].

Example 2.3

The set

$$K = \{y \in H^1(\Omega) \mid 0 \leq y(x) \leq 1 \quad \text{a.e. in} \quad \Omega\} \tag{2.21}$$

is polyhedric [19].

3. Differential Stability of Projection

In this section the form of the conical differential of the projection in Hilbert space onto the polyhedric subset is presented. We derive also the conical differential of the projection onto a convex set of the form (3.3).

Finally the form of the conical differential of the projection for two examples is presented. The first example concerns the convex sets given by a finite number of linear constraints and the second example concerns the so-called convex polyhedra. The following result has been obtained by Mignot [19], who has shown that if the set $K \subset H$ is polyhedric then the mapping (2.6) is conically, differentiable. We refer the reader to Haraux [7] for the related results for the metric projection.

Lemma 3.1

Assume that the set $K \subset H$ is polyhedric. Then for any $f \in H'$ and for $\varepsilon > 0$, ε small enough

$$\forall h \in H' : P_K(f+\varepsilon h) = P_K(f) + \varepsilon P_S(h) + o(\varepsilon) \quad \text{in} \quad H \tag{3.1}$$

where $\|o(\varepsilon)\|_H/\varepsilon \to 0$ with $\varepsilon \downarrow 0$ uniformly with respect to h on compact subsets of H'. Here we denote

$$S = S_K(f) \quad , \quad \forall f \in H'$$

The proof of Lemma 3.1 is given e.g. in [19]. ■

It is usefull for the applications to the sensitivity analysis of optimal control problems to introduce the following variational inequality

$$\begin{aligned} & y = \Pi(f) \in U \\ & b(y,v-y) \geq \langle f,v-y \rangle_{W',W} \\ & \qquad \forall v \in U \end{aligned} \tag{3.2}$$

where W is a Hilbert apece, $U \subset W$ is a closed, convex subset of the form

$$U = \{\phi \in W \mid R\phi \in K \subset H\} \tag{3.3}$$

here $R \in \mathcal{L}(W;H)$ is a given continous, linear mapping and $K \subset H$ is a closed, convex subset of Hilbert space H.

In order to assume the existence and uniqueness of the solution to (3.2) we assure that the bilinear form $b(.,.) : W \times W \to R$ is coercive and continuous i.e., verifies the conditions (2.2), (2.3), furthermore we assume that it is symmetric

$$b(u,v) = b(v,u) \quad , \quad \forall u,v \in W$$

We show that the conical differentiability of the mapping:

$$W' \ni f \longrightarrow \Pi(f) \in U \subset W \tag{3.4}$$

is equivalent to the conical differentiability of a certain metric projection mapping $P(.) : H \to K \subset H$. We assume that operator R maps W onto H and that $0 \in K \subset H$ therefore

$$\ker R \cap U = \ker R \tag{3.5}$$

We denote

$$W_1 = \ker R \, , \quad W_2 = W_1^{\perp} \tag{3.6}$$

thus

$$W = W_1 \oplus W_2 \tag{3.7}$$

and there exists the inverse $R^{-1} \in \mathcal{L}(H;W_2)$. We define the bilinear form $a(.,.) : H \times H \to R$ in the following way:

$$a(h_1,h_2) \overset{\text{def}}{=} b(R^{-1}h_1, R^{-1}h_2) \, , \quad \forall h_1,h_2 \in H \tag{3.8}$$

For a given element $f \in W'$ we denote by $\Phi(f) \in H$ an element given by a unique solution of the variational equation:

$$a(\Phi(f),h = \langle f, R^{-1}h \rangle_{W'W} \, , \quad \forall h \in H \tag{3.9}$$

where $\langle .,. \rangle_{W'W}$ denotes duality pairing between W' and W.

Let us note that the linear mapping

$$W' \ni f \longrightarrow \Phi(f) \in H \tag{3.10}$$

is continuous.
Now we are in position to decompose the solution of the variational inequality (3.2) in the following way:

the solution $y = \Pi(f)$ to (3.2) can be represented in the form

$$\Pi(f) = y_1 + y_2 \, , \quad y_i \in W_i \, , \quad i=1,2 \tag{3.11}$$

where $y_1 \in W_1$ is given by a unique solition of the variational equation:

$$\begin{aligned} & y_1 \in W_1 \\ & b(y_1,\eta) = \langle f,\eta \rangle_{W'W} \, , \quad \forall \eta \in W_1 \end{aligned} \tag{3.12}$$

The element $y_2 \in W_2$ is given by:

$$y_2 = R^{-1}P(\Phi(f)) \tag{3.13}$$

Here the projection operator

$$P : H \longrightarrow K \subset H \tag{3.14}$$

is defined as the mapping, which to a given element $\xi \in H$ assigns a unique solution of the following variational inequality:

$$\begin{aligned} & p = P(\xi) \in K \\ & a(p-\xi, h-p) \geq 0 , \quad \forall h \in K \end{aligned} \tag{3.15}$$

From (3.10), (3.11) and (3.12) we obtain the following Lemma 3.2.

Lemma 3.2

The mapping (3.4) is conically differentiable if and only if the projection operator (3.14) is conically differentiable.

Let us consider an example.

Example 3.1

Let $\phi_i \in W$, for $i=1,\dots,N$ be given elements. Denote by $U \subset W$ the convex, closed set of the form:

$$U = \{v \in W \mid (v,\phi_i)_W \leq b_i , \quad i=1,\dots,N\} \tag{3.16}$$

where $b_i \in R$, $i=1,\dots,N$, are given numbers such that the set (3.16) is nonempty. We will consider the differential stability of the metric projection $P_U(.)$ in W onto U i.e., for any fixed element $f \in W$, the element $y=P_U(f)$ is given by a unique solution of the following variational inequality

$$\begin{aligned} & y = P_U(f) \in U \\ & (y-f, v-y)_W \geq 0 , \quad \forall v \in U \end{aligned} \tag{3.17}$$

Lemma 3.3

The mapping $P_U(.) : W \to U \subset W$ is conically differentiable and for $\varepsilon > 0$, ε small enough

$$\forall h \in H : P_U(f+\varepsilon h) = P_U(f) + \varepsilon Q(h) + o(\varepsilon) \quad \text{in } W$$

where $\| o(\varepsilon)\|_W/\varepsilon \to 0$ with $\varepsilon \downarrow 0$ and $Q(h)=P_S(h)$.

The set $S \subset W$ takes the form:

$$\begin{aligned} S = \{\eta \in W \mid & (\eta,\phi_i)_W \leq 0 \quad \text{for } i \in I_o, \\ & (\eta, y-f)_W = 0\} \end{aligned} \tag{3.18}$$

where

$$I_o = \{i \mid (y,\phi_i)_W = b_i\} \tag{3.19}$$

Proof:

In order to apply Lemma 3.2 to this example we define the bilinear form $b(.,.)$ as follows:

$$b(u,v) \stackrel{def}{=} (u,v)_W \,, \quad \forall u,v \in W \tag{3.20}$$

In this case

$$W_1 = \text{lin}\{\phi_1,\dots,\phi_N\} \tag{3.21}$$

$$W_2 = W_1$$

Let us denote by $\{\psi_i\}_{i=1}^m \subset W_1$ an orthonormal basis in W_1. Thefore for any element $v \in W_1$ there exists the vector $\alpha = \text{col}(\alpha_1,\dots,\alpha_m)$ such that

$$v = \sum_{i=1}^{m} \alpha_i \psi_i \tag{3.22}$$

We denote $H=R^m$, let the linear mapping $R \in \mathcal{L}(W;R^m)$ be defined in the following way

$$\forall \eta \in W : R\eta = \alpha = \text{col}(\alpha_1,\dots,\alpha_m) \tag{3.23}$$

$$\text{where} \quad \alpha_i \stackrel{def}{=} (\eta,\psi_i)_W \,, \quad i=1,\dots,m$$

Denote by $K \subset R^m$ the set of the form (2.19) where

$$\begin{aligned} a^i &= \text{col}(a_1^i,\dots,a_m^i) \\ a_k^i &= (\psi_k,\phi_i)_W \,, \quad i=1,\dots,N \,, \quad k=1,\dots,m \end{aligned} \tag{3.24}$$

Since the set (2.19) is polyhedric therefore the metric projection in R^m onto the set (2.19) is conically differentiable and from Lemma 3.2 it follows that the metric projection in W onto (2.39) is conically differentiable. In particular it follows that the set (2.39) is polyhedric. ∎

Finaly let us consider the differential stability of metric projection onto the so-called convex polyhedra [5].

Example 3.2

Let there be given a sequence $\{\phi_i\}_1^\infty \subset L^2(\Omega)$ such that

$$\int_\Omega \phi_i(x)\phi_j(x)dx = \delta_{ij} = \begin{cases} 1 \;, & i=j \\ 0 \;, & i\neq j \end{cases} \tag{3.25}$$

Denote

$$W = L^2(\Omega)$$
$$U = \{\eta \in L^2(\Omega) \;|\; 0 \le \int_\Omega \eta(x)\phi_i(x)dx \le a_i \;,\; i=1,2,\ldots\} \tag{3.26}$$

where $a_i > 0$ are given numbers for $i=1,2,\ldots$ Let is consider variational inequality:

$$y = \Pi(f) \in U$$

$$\int_\Omega y(x)(v(x)-y(x))dx \ge \int_\Omega f(x)v(x)-y(x))dx \;,\; \forall v \in U \tag{3.27}$$

where $f \in L^2(\Omega)$ is a given element.

In this example we can apply Lemma 3.2 in order to show that the mapping

$$L^2(\Omega) \ni f \longrightarrow \Pi(f) \in L^2(\Omega) \tag{3.28}$$

defined by the variational inequality (3.27) is actually conically differentiable. To this end let us put

$$b(y,z) = \int_\Omega y(x)z(x)dx \;,\quad \forall y,z \in L^2(\Omega)$$

$$H = \ell^2$$

$$K = \{\{\alpha_i\}_1^\infty \subset \ell^2 \;|\; 0 \le \alpha_i \le a_i \;,\; i=1,2,\ldots\} \tag{3.29}$$

$$R\eta = \{\alpha_i\}_1^\infty \;,\quad \forall \eta \in L^2(\Omega) \tag{3.30}$$

where $\alpha_i = \int_\Omega \eta(x)\phi_i(x)dx \;,\; i=1,2,\ldots$

It can be verified [5] that the orthogonal projection in ℓ^2 onto the set (3.29) is conically differentiable, therefore by Lemma 3.2 it follows that the mapping (3.28) is conically differentiable i.e., for $\varepsilon > 0$, ε small enough:

$$\forall h \in L^2(\Omega) \;:\; \Pi(f+\varepsilon h) = \Pi(f)+\varepsilon\Pi'(h)+o(\varepsilon) \tag{3.31}$$

where $\|o(\varepsilon)\|_{L^2(\Omega)}/\varepsilon \to 0$ with $\varepsilon \downarrow 0$.

The mapping $\Pi'(.) : L^2(\Omega) \longrightarrow L^2(\Omega)$ is given by a unique solution of the following variational inequality:

$$\theta = \Pi'(h) \in S$$

$$\int_\Omega \theta(x)(v(x)-\theta(x))dx \geq \int_\Omega h(x)(v(x)-\theta(x))dx \ , \ \forall v \in S \tag{3.32}$$

where

$$S = \{\eta \in L^2(\Omega) \mid \int_\Omega \eta(x)\phi_i(x)dx \geq 0, \ i \in I_o \tag{3.33}$$

$$\int_\Omega \eta(x)\phi_i(x)dx \leq 0, \ i \in I_1$$

$$\int_\Omega (y(x)-f(x))\eta(x)dx = 0\}$$

here we denote

$$I_o = \{i \in \mathcal{N} \mid \alpha_i = \int_\Omega y(x)\phi_i(x)dx = 0\} \tag{3.34}$$

$$I_1 = \{i \in \mathcal{N} \mid \alpha_i = \int_\Omega y(x)\phi_i(x)dx = a_i\} \tag{3.35}$$

4. Sensitivity Analysis of Abstract Constrained Optimization Problems

This section is devoted to the sensitivity analysis of an abstract, constrained optimization problem. Using the results concerning differential stability of metric projection in Hilbert space onto a closed and convex subset we derive the form [24] of the so-called sensitivity coefficient of an optimal solution to the optimization problem with respect to the parameter. We apply the abstract results to an example of optimal control problem. Let H,Y be Hilbert spaces, we identify the dual space H' with H. Let $K \subset H$ be a given closed and convex subset. Assume that for $\varepsilon \in [0,\delta)$ there are given

- linear mappings $L_\varepsilon \in \mathcal{L}(H;Y)$
- functionals $I_\varepsilon(.) : Y \to R$

Consider the following optimization problem:

(P_ε) find an element $u_\varepsilon \in K$ which minimizes the cost functional

$$J_\varepsilon(u)=I_\varepsilon(L_\varepsilon u) + \frac{\alpha}{2} ||u-u_d||_H^2 \tag{4.1}$$

over the set K.

Theorem 4.1. [24]

Assume that the metric projection $P_K(.)$ in H onto K is conically differentiable i.e., for any f H and for $\tau>0$, τ small enough

$$\forall h \in H : P_K(f+\tau h) = P_K(f) + \tau Q(h) + o(\tau) \tag{4.2}$$

where $||o(\tau)||_H/\tau \to 0$ with $\tau \downarrow 0$, $Q(.):H \to H$ is given continuous mapping in general depending on $f \in H$.

Furthermore assume that the following conditions are satisfied

(i) there exists a linear mapping $L'_o \in \mathcal{L}(H;Y)$ such that if there are given elements

$$v_\varepsilon = v_o + \varepsilon v'_o + r(\varepsilon) \quad \text{in } H, \quad \varepsilon \in [0,\delta)$$

where $v_o, v'_o \in H$, $r(\varepsilon)/\varepsilon \to 0$ weakly in H with $\varepsilon \downarrow 0$ then

$$L_\varepsilon v_\varepsilon = L_o v_o + \varepsilon(L'_o v_o + L_o v'_o) + o(\varepsilon) \tag{4.3}$$

where $||o(\varepsilon)||_H/\varepsilon \to 0$ with $\varepsilon \downarrow 0$.

(ii) functional $I(.)$ is twice continuously differentiable and the following conditions are satisfied for $\varepsilon \in [0,\delta)$

$$\|DI_o(L_o v) - DI_o(L_\varepsilon v)\|_{Y'} \le C\, \|L_o v - L_\varepsilon v\|_Y \le C\varepsilon \|v\|_H, \quad \forall v \in H \tag{4.4}$$

$$\|DI_o(y) - DI_\varepsilon(y)\|_{Y'} \le C\varepsilon \|y\|_Y, \quad \forall y \in Y \tag{4.5}$$

where C is a generic constant

(iii) there exists a constant $\sigma > 0$ such that

$$(L_o^* DI_o(L_o v) - L_o^* DI_o(L_o u), v-u)_H + \alpha\, ||v-u||_H^2 \ge \sigma\, ||v-u||_H^2, \quad \forall v,u \in H$$

then for $\varepsilon > 0$, ε small enough

$$u_\varepsilon = u_o + \varepsilon q + o(\varepsilon) \quad \text{in } H \tag{4.7}$$

where $||o(\varepsilon)||_H/\varepsilon \to 0$ with $\varepsilon \downarrow 0$. The sensitivity coefficient $q \in H$ verifies the following conditions

$$q \in S_K(F_o) \tag{4.8}$$

$$q = Q(F'_o) \tag{4.9}$$

where

$$S_K(F_o) = \overline{C_K(P_K(F_o))} \cap T_K(F_o) \tag{4.10}$$

$$F_o = u_d - \frac{1}{\alpha} L_o^* D I_o(L_o u_o) \tag{4.11}$$

$$F_o' = \frac{1}{\alpha} (L_o')^* D I_o(L_o u_o) - \frac{1}{\alpha} L_o^* D^2 I_o(L_o u_o; L_o' u_o - L_o q) \tag{4.12}$$

The proof of Theorem 4.2 is given in [24].

Remark 4.1.:

Let us note that by (4.9), (4.12) the sensitivity coefficient q is the fixed point of a nonlinear mapping.

Corrolary 4.1

Assume that the set $K \subset H$ is polyhedric. Then the sensitivity coefficient q in 4.7 is uniquely determined. The element $q \in H$ is the fixed point of the following nonlinear mapping

$$q = P_{S_K(F_o)}(F_o'(q)) \tag{4.13}$$

■

Let us consider the problem with state constraints in our abstract setting. To this end assume that there are given closed, convex subsets

$$U_{ad} \subset H \tag{4.14}$$

$$Y_{ad} \subset Y \tag{4.15}$$

and that the set $K \subset H$ takes on the form:

$$K = \{u \in H \mid u \in U_{ad}, \ Lu \in Y_{ad}\} \tag{4.16}$$

there we assume that the linear mapping $L \in \mathcal{L}(H;Y)$ does not depends on the parameter.

Denote by $v_\varepsilon \in K$ an element which minimizes the cost functional

$$J_\varepsilon(v) = I_\varepsilon(Lv) + \frac{\alpha}{2} \|v - z_\varepsilon\|_H^2 \tag{4.17}$$

over the set K of the form (4.16).
Here $z_\varepsilon \in H$ are given elements for $\varepsilon \in [0,\delta)$, such that

$$z_\varepsilon = z_o + \varepsilon z' + o(\varepsilon) \quad \text{in} \quad H \tag{4.18}$$

where $\|o(\varepsilon)\|_H/\varepsilon \to 0$ with $\varepsilon \downarrow 0$.

Lemma 4.1

Let the assumptions (ii), (iii) of Theorem 1 be satisfied. Then for $\varepsilon > 0$, ε small enough

$$\| v_\varepsilon - v_o \|_H \leq C\varepsilon \tag{4.19}$$

The proof of Lemma 4.1 follows by Proposition 3 in [24]

Lemma 4.2

Assume that

$$F_o = z_o - \frac{1}{\alpha} L^* DI_o(Lv_o) \in K \tag{4.20}$$

then for $\varepsilon > 0$, ε small enough

$$v_\varepsilon = v_o + \varepsilon q + o(\varepsilon) \quad \text{in} \quad H \tag{4.21}$$

The element $q \quad S = \overline{C_K(F_o)}$ is uniquely determined in the form of the fixed point

$$q = P_S(F_o'(q)) \tag{4.22}$$

where

$$F_o'(q) = \frac{1}{\alpha} LD^2 I_o(Lv_o; Lq) \tag{4.23}$$

The proof of Lemma 4.2 follows by Lemma 2.2 and Theorem 4.1.

■

Let us consider an example of control and state constrained optimal control problem for parabolic equation.

Example 4.1

Let $\Omega \subset R^n$ be a given domain with smooth boundary $\Gamma = \partial\Omega$. Denote $Q = \Omega \times (0,\tau)$, $\Sigma = \partial\Omega \times (0,\tau)$, where $\tau > 0$.
Let $z_\varepsilon \in L^2(Q)$, $\varepsilon \in [0,\delta)$ be given elements such that

$$z_\varepsilon = z_o + \varepsilon z' + o(\varepsilon) \quad \text{in} \quad L^2(Q) \tag{4.24}$$

where

$$\|o(\varepsilon)\|_{L^2(Q)}/\varepsilon \to 0 \quad \text{with} \quad \varepsilon \downarrow 0.$$

In order to define an optimal control problem we introduce the state equation the cost functional and the constraints of the form:

state equation:

$$\frac{\partial y}{\partial t} - \Delta y = 0 \quad \text{in} \quad Q$$

$$y = u \quad \text{on} \quad \Sigma \tag{4.25}$$

$$y(x,0) = 0 \quad \text{on} \quad \Omega$$

where $u(.,.) \in L^2(\Sigma)$ denotes control.

cost functional:

$$J_\varepsilon(u) = \frac{1}{2} \int_Q (y-z_\varepsilon)^2 dQ + \frac{\alpha}{2} \int_\Sigma (u)^2 d\Sigma \tag{4.26}$$

where $\alpha > 0$

constraints:

$$y(.,.) \in Y_{ad} \subset L^2(Q) \tag{4.27}$$

$$u(.,.) \in U_{ad} \subset L^2(\Sigma) \tag{4.28}$$

where Y_{ad}, U_{ad} are closed, convex subsets of $L^2(Q)$, $L^2(\Sigma)$, respectively.

We denote by $K \subset L^2(\Sigma)$ the set of the form:

$$K = \{u \in L^2(\Sigma) \mid u \in U_{ad} \ , \ y=Lu \in Y_{ad}\} \tag{4.29}$$

here we denote $y=Lu$ a unique solution of the state equation (4.25). We assume that the set (4.29) is nonempty.

Remark 4.2

The weak solution of the state equation (4.25) is defined in the following way [11]:

find an element $y \in L^2(Q)$ such that

$$\int_Q y\left(-\frac{\partial z}{\partial t} - \Delta z\right) dQ = \int_\Sigma u \frac{\partial z}{\partial n} d\Sigma \tag{4.30}$$

$$\forall z \in H^{2,1}(Q) \cap L^2(0,T;H_0^1(\Omega)) \ , \ z(x,\tau)=0 \quad \text{on} \quad \Omega$$

∎

It can be verified that for any $\varepsilon \in [0,\delta)$ there exists a unique optimal control $u_\varepsilon \in L^2(\Sigma)$ which minimizes the cost functional (4.26) subject to state equation (4.25) and the constraints (4.27), (4.28).

The optimal control u_ε satisfies the following optimality system which consists of the state equation, the adjoint state equation and the optimality condition.

<u>Optimality system</u>

find $(y_\varepsilon, p_\varepsilon, u_\varepsilon) \in Y_{ad} \times H^{2,1}(Q) \times U_{ad}$ such that

$$\left\{\begin{aligned} &\frac{\partial y_\varepsilon}{\partial t} - \Delta y_\varepsilon = 0 \quad \text{in} \quad Q \\ &y_\varepsilon = u_\varepsilon \quad \text{on} \quad \Sigma \\ &y_\varepsilon(x,0) = 0 \quad \text{on} \quad \Omega \end{aligned}\right. \tag{4.31}$$

$$\left\{\begin{aligned} &\frac{\partial p_\varepsilon}{\partial t} - \Delta p_\varepsilon = y_\varepsilon - z_\varepsilon \quad \text{in} \quad Q \\ &p_\varepsilon = 0 \quad \text{on} \quad \Sigma \\ &p_\varepsilon(x,\tau) = 0 \quad \text{on} \quad \Omega \end{aligned}\right. \tag{4.32}$$

$$u_\varepsilon \in K : \int_\Sigma (\alpha u_\varepsilon - \frac{\partial p_\varepsilon}{\partial n})\,(v-u_\varepsilon)d\Sigma \geq 0, \ \forall v \in K \tag{4.33}$$

■

Let us note that the condition (4.33) says that

$$u_\varepsilon = P_K\left(\frac{1}{\alpha}\frac{\partial p_\varepsilon}{\partial n}\right) \tag{4.34}$$

where the set $K \subset L^2(\Sigma)$ is given by (4.29). From Lemma 4.1 it follows that for $\varepsilon > 0$, ε small enough

$$\|u_\varepsilon - u_o\|_{L^2(\Sigma)} \leq C\varepsilon \tag{4.35}$$

From (4.35) we obtain

$$u_\varepsilon = u_o + \varepsilon q + r(\varepsilon) \quad \text{in} \quad L^2(\Sigma) \tag{4.36}$$

where $r(\varepsilon)/\varepsilon \longrightarrow 0$ weakly in $L^2(\Sigma)$ with $\varepsilon \downarrow 0$. The element $q \in L^2(\Sigma)$ is the so-called sensitivity coefficient of the optimal control u_o. From Lemma 2.1 we have that

$$q \in S_K\left(\frac{1}{\alpha}\frac{\partial p_o}{\partial n}\right) \tag{4.37}$$

We show in some particular cases that the sensitivity coefficient q is uniquely determined in the form of a unique optimal control for an auxiliary optimal control problem.

Lemma 4.3

Assume that

$$\frac{1}{\alpha}\frac{\partial p_o}{\partial n} \in K \tag{4.38}$$

then for $\varepsilon > 0$, ε small enough

$$u_\varepsilon = u_o + \varepsilon q + o(\varepsilon) \quad \text{in} \quad L^2(\Sigma) \tag{4.39}$$

where $\|o(\varepsilon)\|_{L^2(\Sigma)}/\varepsilon \to 0$ with $\varepsilon \downarrow 0$.

The sensitivity coefficient $q \in L^2(\Sigma)$ is given by a unique solution of the following optimal control problem:

find an element $q \in L^2(\Sigma)$ which minimizes the cost functional

$$I(v) = \frac{1}{2}\int_Q (w-z')^2 dQ + \frac{\alpha}{2}\int_\Sigma (v)^2 d\Sigma \tag{4.40}$$

here $z' = \lim_{\varepsilon \downarrow 0} (z_\varepsilon - z_o)/\varepsilon$ in $L^2(Q)$

subject to

state equation:

find an element $w \in L^2(Q)$ such that

$$\int_Q w\left(-\frac{\partial\psi}{\partial t} - \Delta\psi\right)dQ = \int_\Sigma v\frac{\partial\psi}{\partial n}\, d\Sigma \tag{4.41}$$

$$\forall\psi \in H^{2,1}(Q) \cap L^2(0,T;H_o^1(\Omega)), \quad \psi(x,\tau)=0 \quad \text{on} \quad \Omega$$

and

constraints:

$$v \in \overline{C_{U_{ad}}(u_o)} \tag{4.42}$$

$$w \in \overline{C_{Y_{ad}}(y_o)} \tag{4.43}$$

The proof of Lemma 4.3 follows from Theorem 4.1 and Lemma 2.2 and is omitted here.

■

In the sequal we derive the form of the sensitivity coefficient q for the optimal control u_ε in the case of finite number of linear constraints. We assume that the sets of admissible states and controls are given by

$$Y_{ad} = \{\eta \in L^2(Q) \mid \int_Q \eta\phi_i dQ \leq a_i,\ i=1,\dots,M\} \tag{4.44}$$

$$U_{ad} = \{v \in L^2(\Sigma) \mid \int_\Sigma v\psi_i d\Sigma \leq b_i,\ i=1,\dots,N\} \tag{4.45}$$

respectively, where

$$\phi_i \in L^2(Q),\ a_i \in R,\ i=1,\dots,M$$

$$\psi_i \in L^2(\Sigma),\ b_i \in R,\ i=1,\dots,N$$

are given elements.
It can be cerified that in this example the set (4.16) is given by:

$$K = \{v \in L^2(\Sigma) \mid v \in U_{ad},\ \int_\Sigma v \frac{\partial z_i}{\partial n} d\Sigma \leq a_i,\ i=1,\dots,M\} \tag{4.46}$$

where the elements z_i, $i=1,\dots,M$ satisfy the parabolic equation:

$$\begin{aligned} -\frac{\partial z_i}{\partial t} - \Delta z_i &= \phi_i, \quad \text{in } Q \\ z_i &= 0 \quad \text{on } \Sigma \\ z_i(x,\tau) &= 0 \quad \text{on } \Omega \end{aligned} \tag{4.47}$$

Let us note that the conical differentiability of metric projection in $L^2(\Sigma)$ onto the set (4.46) follows by Example 3.1. Actually we have the following Lemma.

Lemma 4.4

The set (4.46) is polyhedric; for any $f \in L^2(\Sigma)$ the cone $S_K(f)$ is given by

$$S_K(f) = \{v \in L^2(\Sigma) \mid \int_\Sigma v\psi_i d\Sigma \leq 0,\ i \in I_o,\ \int_\Sigma v \frac{\partial z_i}{\partial n} d\Sigma \leq 0,\ i \in J_o,\ \int_\Sigma (u-f)v d\Sigma = 0\} \tag{4.48}$$

where the element $u=P_K(f)$ is given by a unique solution of the following variational inequality:

$$u \in K$$
$$\int_\Sigma (u-f)(v-u)d\Sigma \geq 0, \quad \forall v \in K \tag{4.49}$$

Here we denote

$$I_o = \{i \in \{1,\dots,N\} \mid \int_\Sigma u\psi_i d\Sigma = b_i\} \tag{4.50}$$

$$J_o = \{i \in \{1,\dots,M\} \mid \int_\Sigma u \frac{\partial z_i}{\partial n} d\Sigma = a_i\} \tag{4.51}$$

The proof of Lemma 4.4 is omitted here.

Theorem 4.2

A unique optimal control $u_\varepsilon \in L^2(\Sigma)$ for the problem (4.25)÷(4.28), (4.44), (4.45) and for $\varepsilon > 0$, ε small enough takes the form:

$$u_\varepsilon = u_o + \varepsilon q + o(\varepsilon) \text{ in } L^2(\Sigma) \tag{4.52}$$

where $\|o(\varepsilon)\|_{L^2(\Sigma)}/\varepsilon \to 0$ with $\varepsilon \downarrow 0$.

The element $q \in L^2(\Sigma)$ is given by a unique optimal solution of an auxiliary optimal control problem. In order to define the optimal control problem we introduce the constraints of the form:

$$\int_Q w\phi_i dQ \leq 0, \text{ for } i \in J_o \tag{4.53}$$

$$\int_\Sigma v\psi_i d\Sigma \leq 0, \text{ for } i \in I_o \tag{4.54}$$

$$\int_\Sigma (u_o - \frac{1}{\alpha}\frac{\partial p_o}{\partial n}) v d\Sigma = 0 \tag{4.55}$$

here p_o denotes a unique solution to the adjoint state equation (4.32) for $\varepsilon=0$. The sets of indices I_o, J_o are given by

$$I_o = \{i \in \{1,\dots,N\} \mid \int_\Sigma u_o\psi_i d\Sigma = b_i\} \tag{4.56}$$

$$J_o = \{j \in \{1,\dots,M\} \mid \int_Q y_o\phi_i dQ = a_i\} \tag{4.57}$$

The sensitivity coefficient $q \in L^2(\Sigma)$ minimizes the cost functional

(4.40) subject to state equation (4.41) and constraints (4.46)÷(4.48). The proof of Theorem 4.2 follows from the results presented in [24,25] and is omitted here.

5. Sensitivity Analysis of Constrained Optimal Control Problems

In this section further examples of the local sensitivity analysis of optimal solutions of optimal control problems for distributed parameter systems are presented. We will consider control problems with control or state constraints. We restrict ourselves to the particular case of polyhedric admissible sets of controls or states however the method presented here is general [24, 25, 28] provided that the form of the conical differential of the respective projection is known.

5.1. Sensitivity Analysis of Control Constrained Optimal Control Problem

We will consider a control constrained, convex optimal control problem for elliptic equation with distributed control.
Let $\Omega \subset R^n$ be a given domain with smooth boundary $\Gamma = \partial\Omega$. Let Ω_o, $E \subset \Omega$ be given domains such that $\Omega_o \cap E = \phi$.
Consider the following optimal control problem which consists of the state equation, the cost functional and the set of admissible controls of the form:

state equation:

find an element $y \in H_o^1(\Omega)$ such that

$$\int_\Omega a_\varepsilon(x)\, y(u;x).\, \phi(x)dx = \int_\Omega f(x)\phi(x)dx + \int_E u(x)\phi(x)dx \qquad \forall \phi \in H_o^1(\Omega) \tag{5.1}$$

here $u(.) \in L^2(E)$ denotes control, $f(.) \in L^2(\Omega)$ is a given element, $a_\varepsilon(.) \in L^\infty(\Omega)$, $\varepsilon \in [0,\delta)$ are given elements such that $a_\varepsilon(x) \geq c > 0$, $x \in \Omega$, $\varepsilon \in [0,\delta)$.

cost functional:

$$J_\varepsilon(u) = \frac{1}{2}\int_{\Omega_o} |\max\{0, y(u;x) - z_d(x)\}|^2 dx + \frac{\alpha}{2}\int_E (u(x))^2 dx \tag{5.2}$$

$\alpha > 0$, $z_d \in L^2(\Omega_o)$ is given element.

set of admissible controls:

$$K = \{u \in L^2(E) \mid 0 \le u(x) \le 1 \text{ a.e. in } E, \quad \int_E u(x)dx \le M\} \tag{5.3}$$

where $M > 0$ is a given constant.

It can be shown that there exists a unique optimal control $u_\varepsilon \in L^2(E)$ which minimizes the cost functional (5.2) over the set (5.3). The optimal control is given by a unique solution of the following optimality system which consists of state equation, adjoint state equation and optimality condition of the form.

Optimality system

find $(y_\varepsilon, p_\varepsilon, u_\varepsilon) \in H_0^1(\Omega) \times H_0^1(\Omega) \times K$ which satisfy:

state equation:

$$\int_\Omega a_\varepsilon(x)\, y_\varepsilon(x).\, \phi(x)dx = \int_\Omega f(x)\phi(x)dx + \int_E u_\varepsilon(x)\phi(x)dx, \quad \forall \phi \in H_0^1(\Omega) \tag{5.4}$$

adjoint state equation:

$$\int_\Omega a_\varepsilon(x)\, p_\varepsilon(x).\, \phi(x)dx = \int_{\Omega_0} \max\{0, y_\varepsilon(x) - z_d(x)\}\phi(x)dx, \quad \forall \phi \in H_0^1(\Omega) \tag{5.5}$$

optimality condition:

$$\int_E (u_\varepsilon(x) - \alpha p_\varepsilon(x))(v(x) - u_\varepsilon(x))dx \ge 0, \quad \forall v \in K \tag{5.6}$$

■

In order to derive the form of the sensitivity coefficient $q \in L^2(E)$ of the optimal control u with respect to the parameter ε at $\varepsilon = 0^+$ we need the following results.

Lemma 5.1

The set $K \subset L^2(E)$ of the form (5.3) is polyhedric. ■

The proof of Lemma 5.1 is given in [28].

Lemma 5.2

The functional of the form:

$$I(y) = \int_{\Omega_o} |\max\{0, y(x) - z_d(x)\}|^2 dx \ , \quad y \in L^2(E) \tag{5.7}$$

is $C^{1,1}$ i.e., it is continuously differentiable, the gradient $\nabla I(y) \in L^2(E)$ has the form:

$$(\nabla I(y), h)_{L^2(\Omega)} = \int_{\Omega_o} \max\{0, y(x) - z_d(x)\} h(x) dx \ , \quad \forall h \in L^2(\Omega_o) \tag{5.8}$$

The mapping

$$L^2(\Omega_o) \ni y \longrightarrow \nabla I(y) \in L^2(\Omega_o) \tag{5.9}$$

is Lipschitz continuous and directionally differentiable and for any $z \in L^2(\Omega_o)$:

$$\lim_{\tau \downarrow 0} (\nabla I(y+\tau z) - \nabla I(y), h)_{L^2(\Omega_o)} / \tau = \int_{\Omega_o^o} h(x) \max\{0, z(x)\} dx + $$

$$+ \int_{\Omega_o^+} h(x) z(x) dx \tag{5.10}$$

where

$$\Omega_o^o = \{x \in \Omega_o \mid y(x) = z_d(x)\} \tag{5.11}$$

$$\Omega_o^+ = \{x \in \Omega_o \mid y(x) > z_d(x)\} \tag{5.12}$$

■

The proof of Lemma 5.2 is omitted here.
We derive the form of sensitivity coefficient for optimal control $u_\varepsilon \in L^2(E)$, with respect to ε at $\varepsilon = 0^+$.

Theorem 5.1

Assume that

$$a_\varepsilon = a_o + \varepsilon a_1 + o(\varepsilon) \quad \text{in} \quad L^\infty(\Omega) \tag{5.13}$$

where $\|o(\varepsilon)\|_{L^\infty(\Omega)} / \varepsilon \to 0$ with $\varepsilon \downarrow 0$ and elements $a_o, a_1 \in L^\infty(\Omega)$ are given, furthermore

$$a_o(x) \geq c > 0, \quad \text{a.e. in} \quad \Omega \tag{5.14}$$

Then for $\varepsilon > 0$, ε small enough

$$u_\varepsilon = u_o + \varepsilon q + o(\varepsilon) \quad \text{in} \quad L^2(E) \tag{5.15}$$

where $\|o(\varepsilon)\|_{L^2(E)}/\varepsilon \to 0$ with $\varepsilon \downarrow 0$.

The sensitivity coefficient $q \in L^2(E)$ is given by a unique solution of the following optimality system:

Optimality system

find $(w,z,q) \in H_0^1(\Omega) \times H_0^1(\Omega) \times S$ which satisfies

state equation:

$$\int_\Omega a_0(x)\, w(x).\, \phi(x)dx = -\int_\Omega a_1(x)\, y_0(x).\, \phi(x)dx + \int_E q(x)\phi(x)dx\ , \quad \forall\phi \in H_0^1(\Omega) \tag{5.16}$$

adjoint state equation:

$$\int_\Omega a_0(x)\, z(x).\, \phi(x)dx = -\int_\Omega a_1(x)\, p_0(x).\, \phi(x)dx + \tag{5.17}$$

$$+ \int_{\Omega_0^0} \max\{0,z(x)\}\phi(x)dx + \int_{\Omega_0^+} z(x)\phi(x)dx\ , \quad \forall\phi \in H_0^1(\Omega)$$

here we denote

$$\Omega_0^0 = \{x \in \Omega_0 \mid y_0(x) = z_d(x)\} \tag{5.18}$$

$$\Omega_0^+ = \{x \in \Omega_0 \mid y_0(x) > z_d(x)\} \tag{5.19}$$

optimality condition:

$$\int_E (q(x)-\alpha z(x))(v(x)-q(x))dx \geq 0, \quad \forall v \in S \tag{5.20}$$

where the convex, closed cone $S \subset L^2(E)$ takes on the form:

$$S = \{v \in L^2(E) \mid v(x) \geq 0 \text{ a.e. on } E_0, \tag{5.21}$$

$$v(x) \leq 0 \text{ a.e. on } E_1,$$

$$\int_E v(x)dx \leq 0 \text{ if } \int_E u_0(x)dx = M,$$

$$\int_E (u_0(x) - \frac{1}{\alpha}\, p_0(x))v(x)dx = 0\}$$

here we denote

$$E_0 = \{x \in E \mid u_0(x) = 0\} \tag{5.22}$$

$$E_1 = \{x \in E \mid u_0(x) = 1\} \tag{5.23}$$

Proof:

We use exactly the same argument as in the Example 3.2 of [24].
By Proposition 3 in [24] it follows that for $\varepsilon > 0$, ε small enough:

$$\|u_\varepsilon - u_o\|_{L^2(E)} \leq C\varepsilon \tag{5.24}$$

therefore there exists a sequence $\{\varepsilon_n\}$, $\lim_{n\to\infty} \varepsilon_n = 0$ and an element $q \in L^2(E)$ such that

$$u_{\varepsilon_n} = u_o + \varepsilon_n q + r(\varepsilon_n) \tag{5.25}$$

where $r(\varepsilon_n)/\varepsilon_n \longrightarrow 0$ weakly in $L^2(E)$ with $\varepsilon_n \downarrow 0$.
Using (5.25) and (5.13) we obtain from (5.4) and from (5.5):

$$y_{\varepsilon_n} = y_o + \varepsilon_n w + r(\varepsilon_n) \quad \text{in } H_o^1(\Omega) \tag{5.26}$$

$$p_{\varepsilon_n} = p_o + \varepsilon_n z + o(\varepsilon_n) \quad \text{in } H_o^1(\Omega) \tag{5.27}$$

where $r(\varepsilon_n)/\varepsilon_n \longrightarrow 0$ weakly in $H_o^1(\Omega)$,

$$\|o(\varepsilon_n)\|_{H_o^1(\Omega)}/\varepsilon_n \longrightarrow 0 \quad \text{with } \varepsilon_n \downarrow 0.$$

The elements w,z satisfy the equations (5.16), (5.17), respectively.
Then we use the optimality condition (5.6) in the form:

$$u_{\varepsilon_n} = P_K\left(\frac{1}{\alpha} p_{\varepsilon_n}\right) \tag{5.28}$$

and by (5.27), Lemma 3.1 and Lemma 5.1 we get from (5.28)

$$u_{\varepsilon_n} = u_o + \varepsilon_n P_S\left(\frac{1}{\alpha} z\right) + o(\varepsilon_n) \quad \text{in } L^2(E) \tag{5.29}$$

where $\|o(\varepsilon_n)\|_{L^2(E)}/\varepsilon_n \to 0$ with $\varepsilon_n \downarrow 0$.
Therefore by (5.25) and (5.29) it follows that

$$q = P_S\left(\frac{1}{\alpha} z\right) \tag{5.30}$$

and the elements (w,z,q) are determined by the optimality system (5.16), (5.17), (5.20).

The element $q \in L^2(E)$ is uniquely determined since from (5.16), (5.17), (5.20) it follows that the sensitivity coefficient $q \in S$ minimizes the cost functional of the form:

$$I(q) = \frac{1}{2} \int_{\Omega_o^o} |\max\{0,z(x)\}|^2 dx + \frac{1}{2} \int_{\Omega_o^+} (z(x))^2 dx - \tag{5.31}$$

$$- \int_{\Omega} a_1(x)\, p_o(x)\, z(x) dx + \frac{\alpha}{2} \int_E (q(x))^2\, dx$$

over the set of admissible controls (5.21), here $z \in H_o^1(\Omega)$ denotes the weak solution of the state equation (5.16).

■

5.2. Sensitivity Analysis of State Constrained Optimal Control Problem

We present an example of state constrained optimal control problem for partial differential equation of elliptic type. We prove that a unique optimal solution to this problem is Lipschitz continuous with respect to the parameter. The form of the sensitivity coefficient of an optimal control with respect to the parameter for a set of admissible states is derived.

Let $\Omega \subset R^3$ be given domain with smooth boundary $\partial\Omega$. In order to define the optimal control problem we introduce the state equation the cost functional and the set of admissible states of the form:

state equation:

find an element $y \in H_o^1(\Omega)$ such that

$$\begin{aligned} \operatorname{div}(a_\varepsilon(x)\, y(x)) &= u(x) \quad \text{in } \Omega \\ y &= 0 \quad \text{on } \partial\Omega \end{aligned} \tag{5.32}$$

here $u(.) \in L^2(\Omega)$ denotes control,

and $$a_\varepsilon(x) = a_o(x) + \varepsilon a_1(x), \quad x \in \bar{\Omega}, \quad \varepsilon \in [0,\delta) \tag{5.33}$$

where $a_o(.), a_1(.) \in C^1(\bar{\Omega})$ are given elements such that $a_\varepsilon(x) \geq c > 0$ for $x \in \Omega$ and for all $\varepsilon \in [0,\delta)$.

cost functional:

$$J(u) = \frac{1}{2} \int_\Omega (y(x)-y_d(x))^2 dx + \frac{\alpha}{2} \int_\Omega (u(x))^2 dx \tag{5.34}$$

where $\alpha > 0$, $y_d \in L^2(\Omega)$ is a given element.

set of admissible states:

nonempty, closed and convex set $K \subset H \overset{def}{=} H^2(\Omega) \cap H_0^1(\Omega)$.

■

We denote by $u_\varepsilon \in L^2(\Omega)$ a unique element which minimizes the cost functional (5.34) subject to the state constraints.

Lemma 5.3

For $\varepsilon > 0$, ε small enough

$$\|u_\varepsilon - u_0\|_{L^2(\Omega)} \leq C\varepsilon \tag{5.35}$$

Proof:

Denote by $y_\varepsilon \in H$ the solution to the state equation (5.35) corresponding to the optimal control $u_\varepsilon \in L^2(\Omega)$. Obviously we have

$$y_\varepsilon \in K, \quad \forall \varepsilon \in [0,\delta) \tag{5.36}$$

We prove that

$$\|y_\varepsilon - y_0\|_H \leq C\varepsilon \tag{5.37}$$

To this end we denote

$$a_\varepsilon(y,z) \overset{def}{=} \int_\Omega \{y(x)z(x)+\alpha \operatorname{div}(a_\varepsilon(x)\nabla y(x))\operatorname{div}(a_\varepsilon(x)\nabla z(x)\}dx \tag{5.38}$$

$$\forall y,z \in H$$

$$< f,z > \overset{def}{=} \int_\Omega y_d(x)z(x)dx, \quad \forall z \in H \tag{5.39}$$

It can be verified that the element $y_\varepsilon \in H$ satisfies the following variational inequality:

$$\begin{aligned} & y_\varepsilon \in K \\ & a^\varepsilon(y_\varepsilon, z-y_\varepsilon) \geq < f, z-y_\varepsilon >, \quad \forall z \in K \end{aligned} \tag{5.40}$$

By (5.38) and our assumption (5.33) it follows that

$$a^\varepsilon(y,y) \geq \sigma \|y\|_H^2, \quad \sigma > 0, \quad \forall y \in H \tag{5.41}$$

$$|a^\varepsilon(y,z)-a^0(y,z)| \leq C\varepsilon \|y\|_H \|z\|_H, \quad \forall y,z \in H \tag{5.42}$$

therefore by standard argument we obtain

$$\|y_\varepsilon - y_o\|_H \le \frac{C}{\sigma}\varepsilon \|y_o\|_H \tag{5.43}$$

Since

$$u_\varepsilon(x) = \operatorname{div}(a_\varepsilon(x)\nabla y_\varepsilon(x)) \tag{5.44}$$

$$= a_\varepsilon(x)\Delta y_\varepsilon(x) + \nabla a_\varepsilon(x).\nabla y_\varepsilon(x), \quad x \in \Omega$$

hence by (5.43), (5.33) it follows (5.35).

■

Let us recall that for a bounded domain $\Omega \subset R^3$ it follows [11] that $H^2(\Omega) \subset C(\bar{\Omega})$, where the injection is continuous.

We derive the form of the sensitivity coefficient for an optimal control u_ε with respect to the parameter ε, at $\varepsilon=0^+$. We assume that the set K of admissible states is given by

$$K = \{\phi \in H \mid a_i \le \phi(x_i) \le b_i, \quad i=1,\dots,N\} \tag{5.45}$$

where $a_i, b_i \in R$ are given numbers, $x_i \in \Omega$ are given points for $i=1,\dots,N$.

We need the following result, which shows that the set (5.45) is polyhedric.

Lemma 5.4

Let us consider variational inequality:

find an element $y=P_K(f) \in K$ such that

$$a^o(y,v-y) \ge \langle f,v-y \rangle, \quad \forall v \in K \tag{5.46}$$

For $\tau > 0$, τ small enough

$$\forall h \in H : P_K(f+\tau h) = P_K(f) + \tau P_S(h) + o(\tau) \quad \text{in } H \tag{5.47}$$

where $\|o(\tau)\|_H/\varepsilon \to 0$ with $\varepsilon \downarrow 0$.

The cone $S \subset H$ takes the form:

$$S = \{\phi \in H \mid \phi(x_i) \ge 0, \ i \in I_o=\{j \in \{1,\dots,N\} \mid y(x_j)=a_j\} \tag{5.48}$$

$$\phi(x_i) \le 0, \ i \in I_1=\{j \in \{1,\dots,N\} \mid y(x_j)=b_j\}$$

$$a^o(y,\phi) = \langle f,\phi \rangle \}$$

The proof of Lemma 5.4 follows by Example 3.1 and is omitted here.

Theorem 5.2

Assume that the set of admissible states is given by (5.45) then for $\varepsilon > 0$, ε small enough:

$$u_\varepsilon = u_o + \varepsilon q + o(\varepsilon) \quad \text{in} \quad L^2(\Omega) \tag{5.49}$$

where $\|o(\varepsilon)\|_{L^2(\Omega)}/\varepsilon \to 0$ with $\varepsilon \downarrow 0$.

The element $q \in L^2(\Omega)$ is given by a unique optimal solution of the following state constrained optimal control problem:

find an element $q \in L^2(\Omega)$ which minimizes the cost functional

$$I(v) = \frac{1}{2} \int_\Omega (z(x))^2 dx + \frac{\alpha}{2} \int_\Omega (v(x))^2 dx \tag{5.50}$$

subject to state constraints

$$z(x_i) \geq 0, \ i \quad I_o = \{j \in \{1,\dots,N\} \mid y_o(x_j) = a_j\} \tag{5.51}$$

$$z(x_i) \leq 0, \ i \quad I_1 = \{j \in \{1,\dots,N\} \mid y_o(x_j) = b_j\} \tag{5.52}$$

$$a^o(y^o, z) - \int_\Omega y_d(x) z(x) dx = 0 \tag{5.53}$$

here the element $z \in H$ denotes the solution of the state equation:

$$\begin{aligned} &\operatorname{div}(a_o(x)\nabla z(x)) = v(x) - \operatorname{div}(a_1(x)\nabla y_o(x)) \quad \text{in} \quad \Omega \\ &z(x) = 0 \quad \text{on} \quad \partial\Omega \end{aligned} \tag{5.54}$$

Proof:

It follows by Theorem 1 in [26], and by Lemma 5.4 that for $\varepsilon > 0$, ε small enough

$$y_\varepsilon = y_o + \varepsilon z + o(\varepsilon) \quad \text{in} \quad H \tag{5.55}$$

where $\|o(\varepsilon)\|_H/\varepsilon \to 0$ with $\varepsilon \downarrow 0$. The element $z \in H$ is given [26] by a unique solution of the following variational inequality

$$\begin{aligned} &z \in S \\ &a^o(z, v-z) \geq -a'(y_o, v-z), \quad \forall v \in S \end{aligned} \tag{5.56}$$

where

$$a'(y,w) \overset{\text{def}}{=} \alpha \int_\Omega \{\operatorname{div}(a_o(x)\nabla y(x))\operatorname{div}(a_1(x)\nabla w(x)) \tag{5.57}$$

$$+ \operatorname{div}(a_1(x)\nabla y(x))\operatorname{div}(a_o(x)\nabla w(x))\}dx, \quad \forall y, w \in H$$

and the cone $S \subset H$ takes the form (5.48) with the element $y(.)$ replaced by the element $y_o(.)$.
From (5.55) it follows that

$$u_\varepsilon = u_o + \varepsilon q + o(\varepsilon) \quad \text{in} \quad L^2(\Omega) \tag{5.58}$$

where

$$q(x) \stackrel{\text{def}}{\equiv} \operatorname{div}(a_o(x)\nabla z(x)) + \operatorname{div}(a_1(x)\nabla y_o(x)), \text{for } x \in \Omega \tag{5.59}$$

Simple calculations show, that the sensitivity coefficient q is given by a unique solution of the control problem (5.50)÷(5.54). ■

6. Concluding Remarks

In this chapter the differential stability of projection in Hilbert space onto a convex and closed subset is considered. The notion of conical differentiability of the projection is introduced. The form of the conical differential of the projection for several examples is obtained.

The method of sensitivity analysis of the constrained optimal control problems which has been developped in [24, 25, 26, 28] is used here in order to derive the sensitivity coefficient of the optimal control with respect to the parameter. The sensitivity coefficient is given by a unique solution of an auxiliary constrained optimal control problem. The examples of optimal control problems for distributed parameter systems are provided.

The method has been applied to the sensitivity analysis of an optimal control problem arising in air polution control in an urban area. The numerical results obtained will be reported in a forthcoming paper.

References

[1] M.P. Bendsøe, N. Olhoff, J. Sokołowski: Sensitivity analysis of problems of elasticity with unilateral constraints , MAT-Report No. 1984-10, Mathematical Institute, The Technical University of Denmark, Lyngby, 1984, to appear in: Journal of Structural Mechanics.

[2] R. Benedict, J. Sokołowski and J.P. Zolesio: Shape optimization for contact problems , in: System Modelling and Optimization, Proceedings of the 11th IFIP Conference, Kopenhagen, Danmark, ed. P. Thoft-Christensen. Springer Verlag, LNCIS Vol.59 (1984) 790-799.

[3] G. Duvaut and J.L. Lions: Les inequations en mechanique et en physique, Dunod, Paris, 1972.

[4] A.V. Fiacco: Introduction to Sensitivity ans Stability Analysis in Nonlinear Programming, Academic Press, New York, 1983.

[5] A. Frideman: Variational Principles and Free Boundary Problems, J. Wiley and Sons, New York, 1982.

[6] S. Fitzpatrick and R.R. Phelps: Differentiability of the metric projection in Hilbert space, Trans. Amer. Math. Soc. 270 (1982), 483-501.

[7] A. Haraux: How to differentiate the projection on a convex set in Hilbert space. Some applications to variational inequalities J. Math. Soc. Japan, vol. 29, no. 4, 1977, 625-631.

[8] R.B. Holmes: Smoothness of certain metric projections on Hilbert space, Trans. Amer. Math. Soc. 184 (1973), 87-100.

[9] K. Jittorntrun: Solution point differentiability without strict complementarity in nonlinear programming , Mathematical Programming Studies 21, edited by A.V. Fiacco, Amsterdam 1984, 127-138.

[10] J.L. Lions: Controle optimal de systemes gouvernes par des equations aux derivees partielles, Dunod, Paris, 1968.

[11] J.L. Lions and E. Magenes: Problemes aux limites non homogenes et applications, vol. 1, Dunod, Paris, 1968.

[12] J.L. Lions: Perturbations singulieres dans les problemes aux limites et an controle optimal, Lecture Notes in Mathematics, vol. 323, Springer Verlag, Berlin, 1973.

[13] J.L. Lions: Some Methods in the Mathematical Analysis of Systems and Their Control, Gordon and Breach, New York, 1981.

[14] J.L. Lions: Controle des systemes distribues singuliers, Dunod, Paris, 1983.

[15] K. Malanowski: Differential sensitivity of solutions to convex programming problems without strict complementarity assumption , Opracowanie IBS PAN, ZTS 3-4/83, Warszawa, 1983.

[16] K. Malanowski: Differential stability of solutions to convex, control constrained optimal control problems , Applied Mathematics and Optimization, 12 (1984) 1-14.

[17] K. Malanowski: On differentiability with respect to parameter of solutions to convex optimal control problems subject to state space constraints , Applied Mathematics and Optimization, 12,

(1984), 231-245.

[18] K. Malanowski and J. Sokołowski: Sensitivity of solutions to convex control constrained optimal control problems for distributed parameter systems , to be published.

[19] F. Mignot: Controle dans les inequations variationelles elliptiques , J. Functional Analysis, 22, 1976, 130-185.

[20] P. Neittaanmaki, J. Sokołowski and J.P. Zolesio: Nonsmooth shape optimization for elliptic variational inequalities , to appear.

[21] R.T. Rockaffelar: La theorie des sous-gradients et ses applications a l'optimization, les Presses de l'Universite de Montreal, 1978.

[22] J. Sokołowski: Sensitivity analysis for a class of variational inequalities , in: E.J. Haug, J. Cea (eds.): Optimization of Distributed Parameter Structures, Sijthoff & Noordhoff, Netherlands (1981), vol.2, 1600-1609.

[23] J. Sokołowski: Optimal control in coefficients of boundary value problems with unilateral constraints , Bulletin of the Polish Academy of Sciences, Technical Sciences, Vol. 31, No. 1-12, 1983, 71-81.

[24] J. Sokołowski: Differential stability of solutions to constrained optimization problems , INRIA, Rapport de Recherche, No. 312, Rocquencourt, 1984, to appear in: Applied Mathematics and Optimization.

[25] J. Sokołowski: Differential stability of control constrained optimal control problems for distributed parameter systems , to appear in: Proceedings of 2nd International Conference on Control Theory for Distributed Parameter Systems and Applications, Vorau, Austria, 1984, Springer Verlag.

[26] J. Sokołowski: Sensitivity analysis of control constrained optimal control problems for distributed parameter systems , to be published.

[27] J. Sokołowski: Conical differentiability of projection on convex sets - an application to sensitivity analysis of Signorini V.I. , Technical Report, Institute of Mathematics, University of Genoa, 1981.

[28] J. Sokołowski: Sensitivity analysis of Signorini variational inequality , to appear in: Banach Center Publications, vol XIX.

[29] J. Sokołowski: Differential stability of solutions to boundary optimal control problems for parabolic systems , to appear.

[30] J. Sokołowski and J.-P. Zolesio: Shape sensitivity analysis for variational inequalities , in: System Modelling and Optimization, eds. Drenick R.F., Kozin F., LNCIS, Vol. 38, Springer Verlag, New York, 1982.

[31] J. Sokołowski and J.-P. Zolesio: Derivation par rapport au domaine dans les problemes unilateraux , INRIA, Rocquencourt, Rapport de Recherche, No. 132, 1982.

[32] J. Sokołowski and J.-P. Zolesio: Shape sensitivity analysis of unilateral problems , Publication Mathematiques, No. 67, Universite de Nice, 1985.

[33] J. Sokołowski and J.-P. Zolesio: Shape sensitivity analysis of elastic structures , The Danish Center for Applied Mathematics and Mechanics, Report No. 289, 1984.

[34] J. Sokołowski and J.-P. Zolesio: Derivee par rapport au domaine de la solution d'un probleme unilateral , to be published.

[35] J. Sokołowski and J.-P. Zolesio: Differential stability of solutions to unilateral problems , to appear in: Proceedings of Conference on Free Boundary Problems, Maubuisson, France, Pitman.

[36] J. Sokołowski and J.-P. Zolesio: Shape sensitivity analysis of elastic - plastic torsion of convex cylindrical bars , to be published.

[37] E.H. Zarantonello: Projection on convex sets in Hilbert space and spectral theorem, Contributions to Nonlinear Functional Analysis, Publ. No. 27, Math. Res. Center, Univ. Wisconsin, Academic Press, New York, 1971, 237-424.

Chapter 2

SENSITIVITY OF SOLUTIONS TO CONVEX OPTIMAL CONTROL PROBLEMS FOR PARABOLIC EQUATIONS

Kazimierz Malanowski

1. Introduction

In this chapter sensitivity with respect to a vector parameter of solutions to optimal control problems for systems described by parabolic equations is considered.

It is assumed that the control problem is strongly convex and controls are functions of time, not of the space variables. Control constraints are of pointwise type and satisfy some regularity conditions.

The cost functional, the state equation and the constraint functions depend on a vector parameter h.

It is shown that if data of the problem are sufficiently regular functions of the parameter h, then the solutions of the problem are Lipschitz continuous and directionally (one-sidedly) differentiable functions of h. The right - differential is given by the solution of an auxiliary quadratic optimal control problem.

The Lagrange multipliers associated with the initial problem are also Lipschitz continuous and directionally differentiable functions of h.

The global sensitivity results, related to those presented here, can be found in the book by A.L. Dontchev [2] and in the references listed therein. Those results concern mostly systems described by ordinary differential equations. However, the method used here is similar to that in [2] and is based on some important results concerning the Lipschitz continuity of optimization problems due to W.W. Hager [5].

As far as the differential sensitivity is concerned most of the known results concern differentiability of the so called optimal value function which to each value of the parameter assigns the corresponding optimal value of the cost functional (see e.g. [1, 4, 10, 15]).

Only a few papers are devoted to differentiability with respect to the parameters of solutions to optimal control and related problems. One has to mention here the results due to A. Hareaux and F. Mignot [6, 16] concerning differentiability of projection on convex sets in a Hilbert space, with an application to solutions of variational inequalities that depend on a parameter, and sensitivity results for the solutions of optimal control problems due to J. Sokołowski [17-19].

In all these papers differential properties of the operator of projection on a convex set play a crucial role.

The results of J. Sokołowski are very similar to those presented here, however we use another approach.

Namely, in this paper differential sensitivity results for mathematical programming problems due to K. Jittorntrum [7] are applied to pointwise optimality conditions for the analysed control problems. Using those results and taking advantage of the fact that the optimal solutions and the associated Lagrange multipliers are Lipschitz continuous functions of the parameter, the form of right - differentials of these functions is derived.

The presented approach is quite general and can be applied to different convex control problems subject to pointwise control constraints. For the first time it was used in [13] for systems described by ordinary differential equations.

Some used notations:

R^n is an n-dimensional Euclidean space with the inner-product denoted by $\langle .,. \rangle$ and the norm $\| \cdot \| = \langle \cdot , \cdot \rangle^{1/2}$.

If $f(\cdot,\cdot) : R^n \times R^m \to R^1$ is a twice continuously differentiable function then

$$D_u f(u,x),\ D^2_{ux} f(u,x)$$

denote the first and the second Fréchet derivatives with respect to the respective variables at the point (u,x),

$$\delta^+_{u,v} f(u,x) = \lim_{\alpha \downarrow 0} \frac{1}{\alpha} \left(f(u+\alpha v,x) - f(u,x) \right)$$

and

$$\delta^-_{u,v} f(u,x) = \lim_{\alpha \uparrow 0} \frac{1}{\alpha} \left(f(u+\alpha v,x) - f(u,x) \right)$$

denote the right and the left differentials of $f(\cdot,\cdot)$ with respect to u in the direction v, at the point (u,x).

A similar notation is used if $f(\cdot,\cdot)$ is a mapping from one Banach space to another.

The notation of spaces is based on that in the book [12].

Let $[0,T]$ be a fixed interval, and let $\Omega \subset R^n$ be an open domain with a smooth boundary Γ.

Denote $Q = \Omega \times (0,T)$ and $\Sigma = \Gamma \times (0,T)$.

$L^2(0,T)$ is the Hilbert space of functions square integrable on $(0,T)$ with the inner product $(\cdot,\cdot)$ and the norm $\| \cdot \| = (\cdot,\cdot)^{1/2}$.

Similarly, we denote the Hilbert space of functions square integrable on Ω and Q:

$L^2(\Omega)$ - with the inner product $(\cdot,\cdot)_\Omega$ and the norm $\|\cdot\|_{2,\Omega}$

$L^2(Q)$ - with the inner product $((\cdot,\cdot))_Q$ and the norm $\|\cdot\|_{2,Q}$.

$L^2(0,T;X)$ - denotes the space of functions $f(\cdot)$ defined on $(0,T)$ with the range in the space X, supplied with the norm

$$\|f\|_{L^2(0,T;X)} = \left[\int_0^T \|f(t)\|_X^2 dt\right]^{1/2}$$

$H^i(\Omega)$, i=1,2, denote Sobolev spaces of functions defined on Ω and square integrable together with their i-th derivatives, supplied with the usual norms denoted by $\|\cdot\|_{H^i}$.

$W(0,T)$ - denotes the space of functions $f(\cdot)$ defined on Q, such that

$$f \in L^2(0,T;H^2(\Omega)),\ D_x f \in L^2(Q),$$

supplied with the norm

$$\|f\|_W = \left[\|f\|^2_{L^2(0,T;H^2(\Omega))} + \|D_x f\|^2_{2,Q}\right]^{1/2}.$$

Note that the following estimate holds (see [12])

$$\max_{t\in[0,T]} \|f(t)\|_{H^1} \leq c\,\|f\|_W. \tag{1.1}$$

Furthermore, we denote

$$W_o(0,T) = \{f \in W(0,T) \mid f(\sigma,t) = 0 \quad \text{for} \quad (\sigma,t) \in \Sigma\}.$$

c - denotes a generic constant, not necessarily the same in two different places.

2. Parabolic Control Problem

Let $H \subset R^m$ be an open and convex set of vector parameters h. We are going to introduce a family $\{(P_h)\}$ of convex optimal control problems (P_h).

To this end first we are going to define the state equation. Let Ω be a bounded domain locally situated on one side of its boundary Γ, which is an (n-1)-dimensional manifold of class C^∞.

Let T be a fixed time of control.
Denote $Q=\Omega\times(0,T)$ and $\Sigma=\Gamma\times(0,T)$.
For each $h\in H$, consider a parabolic equation

$$D_t y(x,t) - A(h)y(x,t) = g(x,t) \text{ in } Q \tag{2.1}$$

along with the initial condition

$$y(x,0) = y^{o}(x) \text{ in } \Omega \tag{2.1a}$$

and the Dirichlet-type homogeneous boundary condition

$$y(\sigma,t) = 0 \text{ in } \Sigma \tag{2.1b}$$

where

$$A(h)y(x) \overset{\text{def}}{=} \sum_{i,j=1}^{n} D_{x_j}(a_{ij}(x,h)(D_{x_i}y(x)) - a_o(x,h)y(x). \tag{2.2}$$

There exist constants $0<\rho_1<\rho_2<\infty$, independent of h, such that

$$\rho_1 \sum_{i=1}^{n} \xi_i^2 \leqslant \sum_{i,j=1}^{n} a_{ij}(x,h)\xi_i\xi_j \leqslant \rho_2 \sum_{i=1}^{n} \xi_i^2 \quad \forall x\in\Omega,\ \forall \xi_1,\xi_2\in R^1,\ \forall h\in H. \tag{2.3}$$

The functions $a_o(\cdot,h)$, $a_{ij}(\cdot,h)=a_{ji}(\cdot,h)$, $i,j=1,2,\dots,n$ are of class C^∞ and there exists a constant ρ_3, independent of h, such that

$$|a_o(x,h)|, |a_{ij}(x,h)|, |D_{x_k}a_{ij}(x,h)| \leqslant \rho_3, \quad \forall i,j,k=1,2,\dots,n \quad \forall x\in\Omega,\ \forall h\in H. \tag{2.4}$$

It is well known that the problem (2.1) has a unique solution provided that $y^o(\cdot)$ and $g(\cdot,\cdot)$ are regular enough. We shall need the following result:

Lemma 2.1. ([8] p.209, Th.6.1)
Suppose that the conditions (2.3), (2.4) *hold and*

$$y^o\in H^1(\Omega) \tag{2.5a}$$

$$g\in L^2(Q) \tag{2.5b}$$

then the boundary value problem (2.1) *has a unique solution* $y\in W_o(0,T)$ *and*

$$\|y\|_W \leqslant c\,(\|y^o\|_{H^1} + \|g\|_{2,Q}) \tag{2.6}$$

where c *does not depend on* h.

Now we introduce control putting in (2.1)

$$g(x,t;h) = \langle b(x;h),u(t)\rangle = \sum_{k=1}^{K} b^k(x,h)u^k(t) \tag{2.7}$$

where

$$b(\cdot,h)=[b^1(\cdot,h),b^2(\cdot,h),\dots,b^K(\cdot,h)]^T \in L^2(\Omega), \tag{2.7a}$$

$$u(\cdot) = [u^1(\cdot),u^2(\cdot),\dots,u^K(\cdot)]^T \in L^2(0,T). \tag{2.7b}$$

Hence, our state equation becomes

$$D_t y(x,t) - A(h)y(x,t) = \langle b(x,h),u(t)\rangle \quad \text{in } Q, \tag{2.8}$$

$$y(x,0) = y^o(x) \quad \text{in } \Omega, \tag{2.8a}$$

$$y(\sigma,t) = 0 \quad \text{in } \Sigma. \tag{2.8b}$$

We introduce the cost functional

$$F(u,y,h) = F^1(y(T),h) + F^2(u,h) \overset{\text{def}}{=}$$

$$= \int_\Omega f^1(y(x,T),h)dx + \int_o^T f^2(u(t),h)dt \tag{2.9}$$

Finally, we define the set of admissible control

$$\mathcal{U}_h^{ad} = \{u \in L^2(0,T) \mid u(t) \in U_h^{ad} \quad \text{for a.a.} \quad t \in [0,T]\} \tag{2.10}$$

where

$$U_h^{ad} = \{u \in R^K \mid \phi^\ell(u,h) \leqslant 0, \ \ell=1,2,\dots,L\}. \tag{2.10a}$$

Now we can formulate our optimal control problem

(P_h) | find $(u_h,y_h) \in L^2(0,T) \times W(0,T)$ such that

$$F(u_h,y_h,h) = \min_{u \in U_h^{ad}} F(u,y,h)$$

subject to (2.8).

We shall assume that the following conditions hold:

(i) for each $h \in H$, the functions $f^1(\cdot,h)$ and $f^2(\cdot,h)$ are positive, convex and twice continuously differentiable. Moreover, there exists a constant $\beta > 0$, independent of h, such that

$$\langle v,D^2_{uu}f^2(u,h)v\rangle \geqslant \beta|v|^2 \quad \forall u,v \in R^K, \quad \forall h \in H \tag{2.11}$$

(ii) the functions $f^1(\cdot,\cdot)$ and $D_x f^1(\cdot,\cdot)$ are continuously differentiable on $R^1 \times H$; similarly, $f^2(\cdot,\cdot)$ and $D_u f^2(\cdot,\cdot)$ are continuously differentiable on $R^K \times H$;

(iii) for each $h \in H$, the conditions (2.3), (2.4) and (2.7a) hold;

(iv) the functions $a_o(\cdot,\cdot)$, $a_{ij}(\cdot,\cdot)$, $D_x a_{ij}(\cdot,\cdot)$, $i,j=1,2,\ldots,n$ and $b^k(\cdot,\cdot)$, $k=1,2,\ldots,K$, are continuously differentiable on $\Omega \times H$;

(v) for each $h \in H$ the functions $\phi^\ell(\circ,h)$, $\ell=1,2,\ldots,L$ are convex and twice continuously differentiable;

(vi) the functions $\phi^\ell(\cdot,\cdot)$ and $D_u\phi^\ell(\cdot,\cdot)$, $\ell=1,2,\ldots,L$ are continuously differentiable on $R^K \times H$;

(vii) for each $h \in H$ the set of admissible control is non-empty

$$\mathcal{U}_h^{ad} \neq \emptyset \quad \forall h \in H . \tag{2.12}$$

Note that by (i), (iii), (v) and (vii) for each $h \in H$, Problem (P_h) has a unique solution u_h.

Additionally to (i) through (vii), the following constraint regularity condition is assumed to be satisfied:

(viii) there exists a constant $\gamma > 0$, independent of h, such that

$$|D_u\phi^T_{I_h(u)}(u,h)v\,| \geqslant \gamma|v| \tag{2.13}$$

for each $h \in H$, for each $u \in U_h^{ad}$ and for each v of an appropriate dimension,

where $D_u\phi^T_{I_h(u)}(u,h)$ denotes the matrix, whose columns are the gradients of all the constraint functions $\phi^i(\cdot,h)$ binding at u.

Note that the condition (2.13) implies that there exists an element $\bar{u}_h \in R^K$, such that

$$\phi^\ell(\bar{u}_h,h) < 0 \quad , \quad \ell \in I=\{1,2,\ldots,L\}. \tag{2.14}$$

Our purpose is to analyse dependence of solutions u_h to (P_h) on the parameter h.

3. Lipschitz continuity of solutions with respect to parameter

In an usual way we introduce a Lagrangian for (P_h)

$$\bar{L}(\cdot,\cdot;\cdot;\cdot) : L^2(0,T) \times W_o(0,T) \times L^2(Q) \times H \to R^1$$

$$\bar{L}(u,y;p;h) = F(u,y;h) + ((p,D_t y - A(h)y - \langle b(h),u\rangle))_Q . \quad (3.1)$$

It is well known [11] that there exists a unique Lagrange multiplier $p_h \in L^2(Q)$, such that at $(u_h,y_h;p_h)$ the Lagrangian $\bar{L}$ assumes its degenerate saddle point, i.e.

$$\bar{L}(u_h,y_h;p;h)=\bar{L}(u_h,y_h;p_h;h) \leqslant \bar{L}(u,y;p_h;h) \quad (3.2)$$

$$\text{for all} \quad u \in \mathcal{U}_h^{ad} \; ; \; y \in W_o(0,T) \, , \; y(0) = y^o \; ; \; p \in L^2(Q).$$

Condition (3.2) is equivalent to the following differential conditions

$$D_y\bar{L}(u_h,y_h;p_h;h) = 0 \quad (3.3)$$

$$(D_u\bar{L}(u_h,y_h;p_h,h),u-u_h) \geqslant 0 \qquad \forall u \in \mathcal{U}_h^{ad} \quad (3.4)$$

From (3.3) it follows that p_h satisfies the adjoint equation

$$D_t p_h(x,t) + A(h)p_h(x,t) = 0 \qquad \text{in } Q, \quad (3.5)$$

$$p_h(x,T) = -D_y f^1(y_h(x,T),h) \qquad \text{in } \Omega, \quad (3.5a)$$

$$p_h(\sigma,t) = 0 \qquad \text{in } \Sigma. \quad (3.5b)$$

Since $y_h \in W_o(0,T)$, then by (1.1) and by (i), $D_y f^1(y_h(T),h) \in H^1(\Omega)$, and by Lemma 2.1

$$p_h \in W_o(0,T). \quad (3.6)$$

Taking into account the pointwise character of the constraints (2.10) we find that (3.4) is equivalent to the condition

$$\langle D_u f^2(u_h(t),h) + r_h(t), u - u_h(t)\rangle \geqslant 0 \qquad \forall u \in U_h^{ad}$$

$$\text{for a.a.} \quad t \in [0,T] \quad (3.7)$$

where

$$r_h^k(t) = -(p_h(t),b^k(h)) \qquad k=1,2,\dots,K. \quad (3.7a)$$

From (3.7) it follows that if we treat $r_h(t)$ as a parameter, then $u_h(t)$ is given as the unique solution of the following convex programming problem:

$$(CP_h)\quad \left| \begin{array}{l} \text{find} \quad u_h \in R^K \quad \text{such that} \\ \\ f^2(u_h(t),h) + \langle r_h(t), u_h(t) \rangle = \min\limits_{u \in U_h^{ad}} \{ f^2(u,h) + \langle r_h(t), u \rangle \} \end{array} \right. \tag{3.8}$$

For (CP_h) we introduce the Lagrangian

$$l(u,\lambda;h) : R^K \times R^L \times R^m \to R^1$$

$$l(u;\lambda;h) = f^2(u,h) + \langle r_h(t), u \rangle + \langle \phi(u,h), \lambda \rangle \tag{3.9}$$

By (2.13) there exists a unique Lagrange multiplier $\lambda_h(t)$ such that the Lagrangian (3.9) assumes its saddle point at $(u_h(t), \lambda_h(t))$, i.e.

$$l\,(u_h(t);\lambda;h) \leq l(u_h(t);\lambda_h(t);h) \leq l(u;\lambda_h(t);h)$$

$$\forall u \in R^K, \quad \forall \lambda \in R^L, \quad \lambda^\ell \geq 0\,, \quad \ell = 1,2,\ldots,L. \tag{3.10}$$

Condition (3.10) is equivalent to the following Kuhn-Tucker conditions

$$D_u f^2(u_h(t),h) + r_h(t) + D_u \phi^T(u_h(t),h)\lambda_h(t) = 0\,, \tag{3.11}$$

$$\lambda^\ell(t) \geq 0 \qquad \ell = 1,2,\ldots,L\,, \tag{3.11a}$$

$$\langle \lambda_h(t), \phi(u_h(t),h) \rangle = 0\,. \tag{3.11b}$$

It can be easily shown [9] that $\lambda_h(\cdot)$ is meansurable as a function defined on $[0,T]$.

We shall need some properties of solutions to (P_h). Let us start with the following

<u>Lemma 3.1</u>

<u>For any compact set</u> $\mathcal{H} \subset H$ <u>there exists a constant</u> $c > 0$ <u>such that</u>

$$|r_h(t)|,\ |u_h(t)|,\ |\lambda_h(t)| \leq c \qquad \forall t \in [0,T], \forall h \in \mathcal{H} \tag{3.12}$$

<u>and</u>

$$\|y_h\|_W\,,\ \|p_h\|_W \leq c \qquad \forall h \in \mathcal{H} \tag{3.13}$$

<u>Proof</u>

By continuity of the constraint functions ϕ^i for any $h \in H$, there exists an open ball $S(h,\rho(h))$ with its center at h and radius $\rho(h)$ such that

$$\bar{u}_h \in \mathcal{U}_g^{ad} \qquad \forall g \in S(h,\rho(h))$$

where $\bar{u}_h(\cdot)$ is a function defined on $[0,T]$ which for each $t \in [0,T]$ assumes the value $\bar{u}_h(t)$=const satisfying (2.14).
Let us denote by $y_g(\bar{u}_h)$ the solution to (2.8) for $h=g$ and $u=\bar{u}_h$.

Expanding $\bar{L}(\cdot,\cdot,p_g,g)$ into Taylor's series and using (i) and (3.1) through (3.4), we obtain

$$F(\bar{u}_h, y_g(\bar{u}_h);g) = \bar{L}(\bar{u}_h, y_g(\bar{u}_h);p_g;g) \geq \bar{L}(u_g,y_g;p_g;g) +$$

$$+ (D_u\bar{L}(u_g,y_g;p_g;g),\bar{u}_h-u_g)+((D_y\bar{L}(u_g,y_g;p_g;g),y_g(\bar{u}_h)-y_g))_Q + \tfrac{1}{2}\beta\|u_g-\bar{u}_h\|_2^2 \geq$$

$$\geq F(u_g,y_g;g) + \tfrac{1}{2}\beta\|u_g-\bar{u}_h\|_2^2 \qquad \forall g \in S(h,\rho(h)),$$

which by (ii) implies that there exists a constant $c(h)$, such that

$$\|u_g\|_2 \leq c(h) \qquad \forall g \in S(h,\rho(h)). \tag{3.14}$$

By (2.6), (2.7) and (3.14) there exists a constant $c(h)$, such that

$$\|y_g\|_W \leq c(h) \qquad \forall g \in S(h,\rho(h)). \tag{3.15}$$

Hence, by (1.1) as well as by (i) and (ii)

$$\|D_y f^1(y_g(T),g)\|_{H^1} \leq c(h) \qquad \forall g \in S(h,\rho(h)). \tag{3.16}$$

Using (3.5), (3.16) and Lemma 2.1, we obtain

$$\|p_g\|_W \leq c(h) \qquad \forall g \in S(h,\rho(h)), \tag{3.17}$$

which by (1.1) implies

$$\|p_g(t)\|_{H^1} \leq c(h) \qquad \forall t \in [0,T] \qquad \forall g \in S(h,\rho(h)). \tag{3.18}$$

(3.7a) together with (2.7a) and (3.18) yield

$$|r_g^k(t)| \leq c(h) \qquad \forall t \in [0,T] \qquad \forall g \in S(h,\rho(h)). \tag{3.19}$$

Expanding $l(\cdot;\lambda;g)$ into Taylor's series at u_g and using (2.11), (2.14) and (3.11), we obtain

$$f^2(\bar{u}_h,g)+\langle r_g(t),\bar{u}_h\rangle \geq l(\bar{u}_h;\lambda_g(t);g) \geq l(u_g(t);\lambda_g(t);g) + \frac{\beta}{2}|\bar{u}_h-u_g(t)|^2 \geq$$

$$\geq f^2(u_g(t),g) + \langle r_g(t),u_g(t)\rangle + \frac{\beta}{2}\,|u_h - u_g(t)|^2$$

$$\forall g \in S(h,\rho(h)),$$

which by (i) and (ii) implies

$$|u_g(t)| \leq c(h)\ , \quad \forall t \in [0,T]\ , \quad \forall g \in S(h,\rho(h)). \tag{3.20}$$

Finally, taking advantage of (2.13) we obtain from (3.11)

$$|\lambda_g(t)| = |\lambda_{g,I_g(t)}(t)| \leq \frac{1}{\gamma}\ |D_u\phi^T_{I_g(t)}(u_g(t))\lambda_{g,I_g(t)}(t)| \leq$$

$$\leq |D_u f^2(u_g(t),g)| + |r_g(t)|\ , \tag{3.21}$$

where $\phi_{I_g(t)}$ (similarly $\lambda_{g,I_g(t)}$) denotes a subvector of ϕ (respectively λ_g) containing all components

$$\ell \in I_g(t) \stackrel{\text{def}}{=} \{\ell \in \{1,2,\dots,L\}\,|\,\phi^\ell(u_g(t),g) = 0\}. \tag{3.22}$$

Estimate (3.21) together with (3.19) and (3.20) imply

$$|\lambda_g(t)| \leq c(h) \qquad \forall t \in [0,T]\ , \quad \forall g \in S(h,\rho(h)). \tag{3.23}$$

Note that the set of balls $S(h,\rho(h))$ constitutes an open covering of $\mathcal{H}$. By compactness of $\mathcal{H}$, from this covering we can extract a finite subcovering. Hence, by (3.15), (3.17), (3.19), (3.20) and (3.23) there exists a constant c, independent of h, such that

$$\|y_h\|_W,\ \|p_h\|_W \leq c \qquad\qquad \forall h \in \mathcal{H},$$

$$|r_h(t)|,\ |u_h(t)|,\ |\lambda_h(t)| \leq c \qquad \forall t \in [0,T],\ \forall h \in \mathcal{H},$$

which completes the proof of the lemma. □

Note that (CP_h) can be treated as a convex programming problem depending on an $(m+K)$-dimensional vector parameter $[h,r_h(t)]$. Since the conditions (2.11), (2.13) and (3.12) are satisfied, we can apply to (CP_h) the sensitivity results due to W.W. Hager [5]. By Theorem D.1 in [5] we obtain:

Lemma 3.2

For any compact and convex set $\mathcal{H} \subset H$ there exists a constant c, such that

$$|u_2(t)-u_1(t)|, |\lambda_2(t)-\lambda_1(t)| \leq c[|h_2-h_1|+|r_2(t)-r_1(t)|] \quad \forall t \in [0,T],$$

$$\forall h_1, h_2 \in \mathcal{H} \qquad (3.24)$$

where for the sake of simplicity the subscripts 1 and 2 are used instead of h_1, h_2.

Now let us return to Problems (P_h) and introduce the following new Lagrangians for them

$$L(\cdot,\cdot;\cdot,\cdot;\cdot) : L^2(0,T) \times W(0,T) \times L^2(Q) \times L^2(0,T) \times H \to R^1$$

$$L(u,y;p,\lambda;h) = \bar{L}(u,y;p;h) + (\lambda,\phi(u,h)) =$$

$$= F(u,y;h)+((p,D_t y-A(h)y-\langle b(h),u\rangle))_Q+(\lambda,\phi(u,h)). \quad (3.25)$$

The conditions (3.2) and (3.10) imply the following saddle point condition

$$L(u_h,y_h;p,\lambda;h) \leq L(u_h,y_h;p_h,\lambda_h;h) \leq L(u,y;p_h,\lambda_h;h)$$

for all $u \in L^2(0,T)$; $y \in W_o(0,T)$, $y(0)=y^o$; $p \in L^2(Q)$; $\lambda \in L^2(0,T)$,

$$\lambda(t) \geq 0 \qquad (3.26)$$

The saddle point implies the following stationarity conditions

$$D_u L(u_h,y_h;p_h,\lambda_h;h) = 0 \qquad (3.27a)$$

$$D_y L(u_h,y_h;p_h,\lambda_h;h) = 0 \qquad (3.27b)$$

Lemma 3.3

For any $h_1, h_2 \in \mathcal{H}$ the following estimates hold

$$\|y_2-y_1\|_W \leq c[|h_2-h_1|+\|u_2-u_1\|_2] \qquad (3.28a)$$

$$\|p_2-p_1\|_W \leq c[|h_2-h_1|+\|u_2-u_1\|_2] \qquad (3.28b)$$

Proof

Subtracting (2.8) at h_2 and h_1 and denoting $z=y_2-y_1$, we get

$$D_t z(x,t)-A(h_2)z(x,t) = [A(h_2)y_1(x,t)-A(h_1)y_1(x,t)] +$$

$$+ [\langle b(x,h_1),u_1(t)\rangle - \langle b(x,h_2),u_2(t)\rangle] \quad \text{in } Q, \tag{3.29}$$

$$z(x,0) = 0 \quad \text{in } \Omega, \tag{3.29a}$$

$$z(\sigma,t) = 0 \quad \text{in } \Sigma. \tag{3.29b}$$

Since $y_1 \in W(0,T)$, then by (iv) and (2.2)

$$\|A(h_2)y_1 - A(h_1)y_1\|_{2,Q} \leq c|h_2-h_1| \, \|y_1\|_W \leq c|h_2-h_1|.$$

On the other hand, (iv) and (2.7) imply

$$\|\langle b(h_1),u_1\rangle - \langle b(h_2),u_2\rangle\|_{2,Q} \leq c[|h_2-h_1| + \|u_2-u_1\|_2].$$

Hence, applying Lemma 2.1 to (2.29) we get (3.28a). Note that by (3.28a) (1.1), (i) and (ii)

$$\|D_yF^1(y_2,h_2)-D_yF^1(y_1,h_1)\|_{H^1} \leq c[|h_2-h_1| + \|u_2-u_1\|_2]$$

Using this estimate and repeating the same argument as above, but for Equation (3.5) istead of (2.8), we obtain (3.28b).

□

Taking into consideration the definition (3.7a) we obtain from Lemmas 3.1, 3.2 and 3.3:

<u>Corollary 3.1</u>

<u>For any</u> $h_1,h_2 \in \mathcal{H}$ <u>the following estimate holds</u>

$$\|\lambda_2-\lambda_1\|_2 \leq c[|h_2-h_1| + \|u_2-u_1\|_2]. \tag{3.30}$$

Now we are in a position to prove the following

<u>Theorem 3.1</u>

<u>If the conditions</u> (i) <u>through</u> (viii) <u>hold, then for any compact and convex set</u> $\mathcal{H} \subset H$ <u>there exists a constant</u> $c > 0$ <u>such that</u>

$$\|u_2-u_1\|_2,\ \|y_2-y_1\|_W,\ \|p_2-p_1\|_W,\ \|\nu_2-\nu_1\|_2 \leq c|h_2-h_1| \quad \forall h_1,h_2 \in \mathcal{H} \tag{3.31}$$

<u>Proof</u>

Expanding $L(\cdot,\cdot;p_2,\nu_2;h_1)$ into Taylor's series at (u_2,y_2) and using (i) we obtain

$$L(u_1,y_1;p_2,\nu_2;h_1) \geq L(u_2,y_2;p_2,\nu_2;h_1)+(D_uL(u_2,y_2;p_2,\nu_2;h_1),u_1-u_2)+$$

$$+((D_yL(u_2,y_2;p_2,\nu_2;h_1),y_1-y_2))_Q+\frac{\beta}{2}\|u_1-u_2\|_2^2 \tag{3.32}$$

From (3.26) we get

$$L(u_1,y_1;p_2,\nu_2;h_1) \leq L(u_2,y_2;p_1,\nu_1;h_1) \tag{3.33}$$

Substituting (3.33) into (3.32) yields

$$\|u_2-u_1\|^2 \leq \{[L(u_2,y_2;p_1,\nu_1;h_1) - L(u_2,y_2;p_2,\nu_2;h_1)] +$$

$$+(D_uL(u_2,y_2;p_2,\nu_2;h_1),u_2-u_1)+$$

$$+((D_yL(u_2,y_2;p_2,\nu_2;h_1),y_2-y_1))_Q\}. \tag{3.34}$$

We shall estimate all three terms on the right-hand side of (3.34). Using the definition (3.25) we obtain

$$L(u_2,y_2;p_1,\lambda_1;h_1)-L(u_2,y_2;p_2,\lambda_2;h_1) =$$

$$=((p_1-p_2,D_ty_2-A(h_1)y_2-\langle b(h_1),u_2\rangle))_Q+(\lambda_1-\lambda_2,\phi(u_2,h_1)).$$

On the other hand (u_2,y_2) satisfy the state equation (2.8), i.e.

$$D_ty_2-A(h_2)y_2 = \langle b(h_2),u\rangle,$$

while by (3.11)

$$(\lambda_1-\lambda_2,\phi(u_2,h_2)) \leq 0.$$

Therefore, using (iv), (vi), (2.2), (3.12), (3.13), (3.28) and (3.30), we get

$$L(u_2,y_2;p_1,\lambda_1;h_1)-L(u_2,y_2;p_2,\lambda_2;h_1) \leq ((p_2-p_1,(A(h_1)-A(h_2))y_2))_Q+$$

$$+((p_2-p_1,\langle b(h_1)-b(h_2),u_2\rangle))_Q+(\lambda_2-\lambda_1,\phi(u_2,h_2)-\phi(u_2,h_1)) \leq$$

$$\leq c|h_2-h_1|\,\|y_2\|_W\,\|p_2-p_1\|_{2,Q}+c|h_2-h_1|\,\|u_2\|_2\,\|p_2-p_1\|_{2,Q}+$$

$$+c|h_2-h_1|\,\|\lambda_2-\lambda_1\|_2 \leq c|h_2-h_1|[|h_2-h_1|+\|u_2-u_1\|_2]. \tag{3.35}$$

Using (ii), (iv), (vi), (3.12) and (3.13), we obtain

$$(D_uL(u_2,y_2;p_2,\lambda_2;h_1),u_2-u_1) =$$

$$= (D_uL(u_2,y_2;p_2,\lambda_2;h_1)-D_uL(u_2,y_2;p_2,\lambda_2;h_2),u_2-u_1) =$$

$$= (D_uF^2(u_2,h_1)-D_uF^2(u_2,h_2),u_2-u_1)+((\langle b_2(h_1)-b_2(h_2),u_2-u_1\rangle,p_2))_Q +$$

$$+ (D_u\phi^T(u_2,h_1)-D_u\phi^T(u_2,h_2),u_2-u_1) \leq$$

$$\leq c|h_2-h_1|\ \|u_2-u_1\|_2+c|h_2-h_1|\ \|p_2\|_{2,Q}\|u_2-u_1\|_2+c|h_2-h_1|\ \|u_2-u_1\|_2 \leq$$

$$\leq c|h_2-h_1|\ \|u_2-u_1\|_2. \tag{3.36}$$

Similarly, by (ii), (iv), (3.13) and (3.28), we get

$$((D_yL(u_2,y_2;p_2,\lambda_2;h_1),y_2-y_1)) =$$

$$= ((D_yL(u_2,y_2;p_2,\lambda_2;h_1)-D_yL(u_2,y_2;p_2,\lambda_2;h_2),y_2-y_1)) =$$

$$= (D_yF^1(y_2(T),h_1)-D_yF^1(y_2(T),h_2),y_2(T)-y_1(T))_\Omega+$$

$$+((A(h_2)-A(h_1))p_2,y_2-y_1))_Q \leq c|h_2-h_1|\ \|y_2(T)-y_1(T)\|_{2,\Omega}+$$

$$+c|h_2-h_1|\ \|p_2\|_W\ \|y_2-y_1\|_{2,Q} \leq c|h_2-h_1|[|h_2-h_1|+\|u_2-u_1\|_2]. \tag{3.37}$$

Substituting (3.35)-(3.37) into (3.34) yields

$$\|u_2-u_1\|_2 \leq c|h_2-h_1|,$$

which together with (3.28) and (3.30) completes the proof of the theorem. □

4. Right-differentiability of solutions and of Lagrange multipliers

We are going to prove existence and to find the form of the right-differentials of u_h,y_h,p_h and λ_h at any $h\in H$ in each direction $g\in R^m$, $|g|=1$, i.e. we are looking for

$$\delta_h^+u_{h,g} = \lim_{\alpha\downarrow 0} \frac{1}{\alpha}(u_{h+\alpha g}-u_h) \tag{4.1a}$$

$$\delta_h^+y_{h,g} = \lim_{\alpha\downarrow 0} \frac{1}{\alpha}(y_{h+\alpha g}-y_h) \tag{4.1b}$$

$$\delta_h^+ p_{h,g} = \lim_{\alpha \downarrow 0} \frac{1}{\alpha} (p_{h+\alpha g} - p_h) \tag{4.1c}$$

$$\delta_h^+ \lambda_{h,g} = \lim_{\alpha \downarrow 0} \frac{1}{\alpha} (\lambda_{h+\alpha g} - \lambda_h) \tag{4.1d}$$

where the limits are taken in strong topologies of the respective spaces.

We shall use the same method which is applied in [13] for control problems of systems described by ordinary differential equations.

Note that by (3.31)

$$\|\frac{1}{\alpha} (p_{h+\alpha g} - p_h)\|_W \leq c|g| = c. \tag{4.2}$$

Hence, from any sequence $\{\alpha\} \downarrow 0$ we can extract a subsequence $\{\alpha'\} \subset \{\alpha\}$, such that

$$\frac{1}{\alpha'} (p_{h+\alpha' g} - p_h) \xrightarrow[\alpha' \downarrow 0]{} q_h \tag{4.3}$$

weakly in $W_o(0,T)$. Since the embedding $W(0,T) \subset L^2(Q)$ is compact (see [12]), (4.3) implies

$$\frac{1}{\alpha'} (p_{h+\alpha' g} - p_h) \xrightarrow[\alpha' \downarrow 0]{} q_h \tag{4.4}$$

strongly in $L^2(Q)$.

Form (iv), (3.7a) and (4.4) it follows that

$$\frac{1}{\alpha'} (r^k_{h+\alpha' g} - r^k_h) \xrightarrow[\alpha' \downarrow 0]{} -(q_h, b^k(h)) - (p_h, D_h b^k(h) g) \overset{\text{def}}{=} s^k_h, \quad k=1,2,\ldots,K \tag{4.5}$$

strongly in $L^2(0,T)$.

In particular (4.5) implies that

$$\frac{1}{\alpha'} (r_{h+\alpha' g}(t) - r_h(t)) \longrightarrow s_h(t) \quad \text{for a.a. } t \in [0,T] \tag{4.6}$$

Now let us return to the convex programming problem (CP_h). This time we shall treat it as a parametric programming problem with a parameter $h \in H$.

Hence, taking into account (i), (vii) and (4.6), we can apply to (CP_h) the sensitivity results for mathematical programming problems due to K. Jittorntrum [7]. By Theorem 4 in [7] we find that there exist the limits

$$\lim_{\alpha' \downarrow 0} \frac{1}{\alpha'} (u_{h+\alpha' g}(t) - u_h(t)) = v_h(t), \tag{4.7a}$$

$$\lim_{\alpha' \downarrow 0} \frac{1}{\alpha'}(\lambda_{h+\alpha' g}(t)-\lambda_h(t)) = \mu_h(t) \tag{4.7b}$$

where $v_h(t)$ is given as a solution to the following quadratic programming problem

(QP) | find $v_h(t) \in R^K$, such that

$$k(v_h(t),t) = \min_{v \in V_h^{ad}(t)} \{k(v,t) \overset{\text{def}}{=} \frac{1}{2} < v, M(t)v > + < m(t), v > \} \tag{4.8}$$

where

$$M(t) = D_{uu}^2 f^2(u_h(t),h) + \sum_{\ell=1}^{L} \lambda_h^{\ell}(t) D_{uu}^2 \phi^{\ell}(u_h(t),h) , \tag{4.9a}$$

$$m(t) = D_{uh}^2 f^2(u_h(t),h)g \quad + s_h(t) + \sum_{\ell=1}^{L} \lambda_h^{\ell} D_{uh}^2 \phi^{\ell}(u_h(t),h)g \tag{4.9b}$$

$$V_h^{ad}(t) = \{v \in R^K \,|< D_u\phi^{\ell}(u_h(t),h), v > + < D_h\phi^{\ell}(u_h(t),h), g > = 0 \quad \text{for} \quad \ell \in I_h^C(t)$$

$$< D_u\phi^{\ell}(u_h(t),h), v > + < D_h\phi^{\ell}(u_h(t),h), g > \leq 0 \quad \text{for} \quad \ell \in I_h(t) \setminus I_h^C(t)\}, \tag{4.10}$$

$$I_h(t) = \{\ell \in \{1,2,\ldots,L\} \,|\, \phi^{\ell}(u_h(t),h) = 0\}, \tag{4.11a}$$

$$I_h^C(t) = \{\ell \in I_h(t) \mid \lambda_h^{\ell}(t) > 0 \}. \tag{4.11b}$$

For $\ell \in I_h(t)$, $\mu_h^{\ell}(t)$ are given as the Lagrange multipliers associated with (QP), while

$$\mu_h^{\ell}(t) = 0 \qquad \text{for} \qquad \ell \notin I_h(t). \tag{4.12}$$

Note that by (i), (v) and (3.11a) the matrix $M(t)$ is positive definite, and by (vii) the set $V_h^{ad}(t)$ is non-empty, hence (QP) has a unique solution, and the associated Lagrange multipliers are defined uniquely.

Note that by (3.31) we have

$$\|\frac{1}{\alpha'}(u_{h+\alpha' g}-u_h)\|_2 , \|\frac{1}{\alpha'}(\lambda_{h+\alpha' g}-\lambda_h)\|_2 \leq c . \tag{4.13}$$

By the Lebesgue dominated convergence theorem (see [3] p.151), the convergence (4.7) almost everywhere on $[0,T]$ together with the estimates (4.13) imply

$$\frac{1}{\alpha'}(u_{h+\alpha' g}-u_h) \longrightarrow v_h \tag{4.14a}$$

$$\frac{1}{\alpha'}(\lambda_{h+\alpha' g} - \lambda_h) \longrightarrow \mu_h \tag{4.14b}$$

strongly in $L^2(0,T)$.

Using the state equation (2.8) in the same way as in the proof of Lemma 3.3, and taking advantage of (iv) and (4.14a), we find that

$$\frac{1}{\alpha'}(y_{h+\alpha' g} - y_h) \longrightarrow z_h \tag{4.15}$$

strongly in $W_o(0,T)$, where

$$D_t z_h(x,t) - A(h) z_h(x,t) = (D_h A(h), g) y_h(x,t) + \langle D_h b(x,h) g, u_o \rangle + \langle b(x,h), v_h \rangle \quad \text{in } Q, \tag{4.16}$$

$$z_h(x,0) = 0 \quad \text{in } \Omega, \tag{4.16a}$$

$$z_h(\sigma,t) = 0 \quad \text{in } \Sigma. \tag{4.16b}$$

We denote here

$$(D_h A(h), g) y(x) = \sum_{i,j=1}^{n} D_{x_j}(\langle D_h a_{ij}(x,h), g \rangle D_{x_i} y(x)). \tag{4.17}$$

Similarly using the adjoint equation (3.5) as well as (4.15) we find that

$$\frac{1}{\alpha'}(p_{h+\alpha' g} - p_h) \longrightarrow q_h \tag{4.18}$$

strongly in $W_o(0,T)$, where

$$D_t q_h(x,t) + A(h) q_h(x,t) + (D_h A(h), g) p_h(x,t) = 0 \tag{4.19}$$

$$q_h(x,T) = -D^2_{yy} f^1(y_h(x,T),h) z_h(x,T) - D^2_{yh} f^1(y_h(x,T),h) g \tag{4.19a}$$

$$q_h(\sigma,t) = 0 \tag{4.19b}$$

It is obvious that the elements q_h given by (4.4) and (4.18) coincide. Analysis of the conditions (4.8), (4.16) and (4.19) together with (4.5) shows that the pair (v_h, z_h) is given as a solution of the following quadratic optimal control problem

(QC) find $(v_h, z_h) \in L^2(0,T) \times W(0,T)$ such that

$$G(v_h, z_h) = \min_{v \in \mathcal{V}_h^{ad}} \{ G(v,z) \overset{\text{def}}{=} \frac{1}{2}[(z(T), N_1 z(T))_\Omega + (v, N_2 v)] + [((n_1, z)) + (n_2, z(T))_\Omega + (n_3, v)] \} \tag{4.20}$$

| subject to (4.16)

where

$$N_1 = D^2_{yy}F^1(y_h,h) \tag{4.21a}$$

$$N_2 = D^2_{uu}F^2(u_h,h) + \sum_{\ell=1}^{L} \lambda^{\ell}_h D^2_{uu}\phi^{\ell}(u_h,h) \tag{4.21b}$$

$$n_1 = -(D_hA(h),g)p_h \tag{4.21c}$$

$$n_2 = D^2_{yh}F^1(y_h(T),h)g \tag{4.21d}$$

$$n_3 = D^2_{uh}F^2(u_h,h)g+\bar{s}_h + \sum_{\ell=1}^{L} \lambda^{\ell}_h D^2_{uh}\phi^{\ell}(u_h,h)g \tag{4.21e}$$

$$\bar{s}^k_h = -(p_h,D_hb^k(h)g) \quad k=1,2,\dots,K \tag{4.21f}$$

$$\mathcal{V}^{ad}_h = \{v \in L^2(0,T) \mid v(t) \in V^{ad}_h(t) \quad \text{for a.a. } t \in [0,T]\} \tag{4.21g}$$

It is easy to see that due to (i),(v), (vii) and (3.11a) Problem (QC) has a unique solution, hence the associated multiplier q_h is unique by (4.19), while the multipliers μ^k associated with the control constraints are unique since their values $\mu^k(t)$ are the unique multipliers for (QP).

This shows that the limits v_h, z_h, r_h and μ_h in (4.14), (4.15) and (4.18) are independent of the choice of the sequences $\{\alpha\}$ and $\{\alpha'\}$, therefore they are the right-differentials of the respective functions.

In this way we arrive at the following principal result:

Theorem 4.1

If the assumptions (i) through (viii) hold, then the solutions (u_h,y_h) of (P_h) and the associated multipliers p_h,λ_h are right-differentiable functions at any $h \in H$ in any direction $g \in R^m$, and the respective right-differentials $\delta^+_h u_{h,g}=v_h$, $\delta^+_h y_{h,g}=z_h$, $\delta^+_h p_{h,g}=q_h$, $\delta^+_h \lambda_{h,g}=\mu_h$ are given by the unique solution and by the associated multipliers for the quadratic optimal control problem (QC), and by (4.12).

Note that the optimality conditions for (QC) can be expressed, analogously to (3.3), (3.4), in terms of the Lagrangian $L(u,y;p,\lambda;h)$ in the following simple form

$$D^2_{yy}L(u_h,y_h;p_h,\lambda_h;h)z_h+D^2_{yp}L(u_h,y_h;p_h,\lambda_h;h)q_h+D^2_{yh}L(u_h,y_h;p_h,\lambda_h;h)g = 0, \tag{4.23}$$

$$(D^2_{uu}L(u_h,y_h;p_h,\lambda_h;h)v_h+D^2_{up}L(u_h,y_h;p_h,\lambda_h;h)r_h +$$

$$+ D^2_{uh}L(u_h,y_h;p_h,\lambda_h;h)g_h, v-v_h) \geq 0 \qquad \forall v \in \Upsilon^{ad}. \tag{4.24}$$

5. Continuous differentiability

Using the same argument as in the proof of Theorem 4.1 we find that the left-differentials of u_h and y_h at $h \in H$ in the direction $g \in R^m$

$$v_h^- = \delta^-_{h,g} u_h = \lim_{\alpha \downarrow 0} \frac{1}{\alpha} (u_{h+\alpha g} - u_h)$$

$$z_h^- = \delta^-_{h,g} y_h = \lim_{\alpha \downarrow 0} \frac{1}{\alpha} (y_{h+\alpha g} - y_h)$$

exist and are given as a unique solution of the following quadratic optimal control problem

$$(QC^-) \left| \begin{array}{l} \text{find } (v_h^-, z_h^-) \in L^2(0,T) \times W(0,T) \text{ such that} \\ G(v_h^-, z_h^-) = \min\limits_{v \in (\Upsilon_h^{ad})^-} G(v,z) \\ \text{subject to (4.16),} \end{array} \right. \tag{5.1}$$

where

$$(\Upsilon_h^{ad})^- = \{v \in L^2(0,T) \mid v(t) \in (V_h^{ad}(t))^-\}, \tag{5.2a}$$

$$(V_h^{ad}(t))^- = \{v \in R^K \mid \langle D_u\phi^\ell(u_h(t),h), v\rangle + \langle D_h\phi^\ell(u_h(t),h), g\rangle = 0 \quad \text{for } \ell \in I_h^c(t),$$

$$\langle D_u\phi^\ell(u_h(t),h), v\rangle + \langle D_h\phi^\ell(u_h(t),h), g\rangle \geq 0 \quad \text{for } \ell \in I_h(t) \setminus I_h^c(t)\}. \tag{5.2b}$$

It is easy to see that the solutions to (QC) and (QC^-) are in general different, hence u_h and y_h are not continuously Gâteaux differentiable at h in the direction g.

However, if

$$\text{meas}\{t \in [0,T] \mid I_h(t) \setminus I_h^c(t) \neq \emptyset\} = 0 \tag{5.3}$$

then

$$\delta^+_{h,g} u_h = \delta^-_{h,g} u_h, \quad \delta^+_{h,g} y_h = \delta^-_{h,g} y_h$$

for any direction $g \in R^m$.

Thus, we obtain the following

Corollary 5.1

If the conditions (i) through (viii), as well as (5.3) hold, then the functions u_h, y_h, p_h, λ_h are continuously Gâteaux differentiable at h in any direction $g \in R^m$.

The assumption (5.3) is not necessary for the continuous differentiability of the so called optimal value function

$$F^o(\circ) : H \to R^1 ,$$

which is defined by

$$F^o(h) \overset{\text{def}}{=} F(u_h, y_h, h). \tag{5.4}$$

Indeed, since by (2.8) and (3.11b) for any $h \in H$ the second and the third terms in the Lagrangian (3.25) vanish at $(u_h, y_h, p_h, \lambda_h)$, then we obtain

$$F^o(h) = L(u_h, y_h; p_h, \lambda_h; h). \tag{5.5}$$

Hence

$$\begin{aligned}\delta^+_{h,g}F^o(h) &= (D_u L(u_h, y_h; p_h, \lambda_h; h), v_h) + ((D_y L(u_h, y_h; p_h, \lambda_h; h), z_h))_Q + \\ &+ ((D_p L(u_h, y_h; p_h, \lambda_h; h), q_h))_Q + (D_\lambda L(u_h, y_h; p_h, \lambda_h; h), \mu_h) + \\ &+ \langle D_h L(u_h, y_h; p_h, \lambda_h; h), g \rangle .\end{aligned} \tag{5.6}$$

Note that by (2.8) and (3.25)

$$((D_p L(u_h, y_h; p_h, \lambda_h; h), q_h))_Q = 0 , \tag{5.7}$$

while by (3.25) and (4.12)

$$(D_\lambda L(u_h, y_h; p_h, \lambda_h; h), \mu_h) = 0 . \tag{5.8}$$

Substituting (3.27), (5.7) and (5.8) into (5.6), we obtain

$$\delta^+_{h,g}F^o(h) = \langle D_h L(u_h, y_h; p_h, \lambda_h; h), g \rangle . \tag{5.9a}$$

Similarly, for the left-derivative we get

$$\delta^-_{h,g}F^o(h) = \langle D_h L(u_h, y_h; p_h, \lambda_h; h), g \rangle . \tag{5.9b}$$

Note that by (ii), (iv), (vi) as well as by Theorem 3.1 $D_hL(u_h,y_h;p_h,\lambda_h;h)$ is a continuous function of h. Hence (5.9) implies

Corollary 5.2

If the conditions (i) through (viii) hold, then the optimal value function $F^o(\cdot)$ is continuously differentiable at any $h \in H$ and

$$D_hF^o(h) = D_hL(u_h,y_h;p_h,\lambda_h;h) \tag{5.10}$$

Note that the results of type (5.10) are well known in sensitivity analysis of optimal control problems (see e.g. [1, 4, 10, 15]).

6. Concluding remarks

It is shown that solutions of the considered optimal control problems are Lipschitz continuous and right-differentiable functions of the vector parameter. The right-differential is given as a unique solution of an auxiliary quadratic control problem. If the conditions (5.3) of strict compementarity type hold, then the right-differentials become the continuous Gâteaux differentials.

In deriving these results two assumptions play a crucial role:

- the strong convexity of the cost functional with respect to the control variable (assumption (i)),
- a pointwise character of the control constraints which satisfy the regularity condition (viii).

If conditions of these types hold, then results similar to those derived here can be obtained for other control constrained problems, e.g. for ordinary differential equations (see [13]).

A general abstract approach for this class of problems is to be presented in a forthcomming paper by J. Sokołowski and the author [14].

References

[1] F.H. Clarke: Optimization and Nonsmooth Analysis, J. Wiley and Sons, New York, 1983.

[2] A.L. Dontchev: Perturbations, Approximations and Sensitivity Analysis of Optimal Control Systems. Lecture Notes in Control and Information Sciences, Vol.52, Springer, Berlin, Heidelberg, New York, Tokyo, 1983.

[3] N. Dunford, J.T. Schwartz: Linear Operators. Part I, Interscience Publishers Inc., New York, 1958.

[4] B. Gollan: On the Optimal Value Function of Optimal Control Problems, Zeitschrift fur Analysis und ihre Ahwendingen, 1 (1982), 17-33.

[5] W.W. Hager: Lipschitz Continuity for Constrained Processes. SIAM J. Control 17 (1979) 321-337.

[6] A. Haraux: Dérivation Dans les Inéquations Variationelles, CR Acad. Sci., Serie A, 278 (1974) 1257-1260.

[7] K. Jittorntrum: Solution Point Differentiability without Strict Complementarity in Nonlinear Programming, in: Mathematical Programming Study 21, A.V. Fiacco (Ed.), Amsterdam 1984, pp. 127-138.

[8] O.A. Ladyženskaja, V.A. Solonnikov, N.N. Uraltzeva: Linear and Quasilinear Equations of Parabolic Type, Nauka, Moscow, 1967 (in Russian).

[9] I. Lasiecka, K. Malanowski: On Regularity of Solutions to Convex Optimal Control Problems with Control Constraints for Parabolic Systems, Contr. and Cybern. 6 (1977), No.3-4, 57-74.

[10] F. Lempio, H. Maurer: Differential Stability in Infinite-Dimensional Nonlinear Programming, Appl. Math. Optim. 6 (1980) 139-152.

[11] J.L. Lions: Contrôl Optimal de Systèmes Gouvernés par des Équations aux Dérivées Partielles, Dunod, Paris, 1968.

[12] J.L. Lions, E. Magenes: Problemes aux Limites Non Homogenes et Applicationes, Vol.1, Dunod, Paris, 1968.

[13] K. Malanowski: Differential Stability of Solutions to Convex, Control Constrained Optimal Control Problems, Appl. Math. Optim. 12 (1984) 1-14.

[14] K. Malanowski, J. Sokołowski: Sensitivity of Solutions to Convex, Control Constrained Optimal Control Problems for Distributed Parameter Systems (to be published).

[15] H. Maurer: Differential Stability in Optimal Control Problems, Appl. Math. Optim. 5 (1979), 283-295.

[16] F. Mignot: Contrôle dans les Inéqualitions Variationelles, J. Functional Anal. 22 (1976), 130-185.

[17] J. Sokołowski: Sensitivity Analysis of Control Constrained Optimal Control Problems for Distributed Systems (to be published).

[18] J. Sokołowski: Differential Stability of Solutions to Constrained Optimization Problems, Rapport de Recherche No. 312, Institut National de Recherche en Informatique et en Automatique, Rocquencourt, 1984.

[19] J. Sokołowski: Differential Stability of Control Constrained Optimal Control Problems for Distributed Parameter Systems, in: Proceedings of 2nd International Conference on Control Theory for Distributed Parameter Systems and Applications, Vorau, Austria, July 9-14, 1984 (to be published).

Chapter 3

PARAMETRIC OPTIMIZATION PROBLEMS FOR EVOLUTION INITIAL - BOUNDARY VALUE PROBLEMS

Jan Sokołowski

1. Introduction

This chapter is concerned with the parametric optimization problems for evolution equations. Using the concept of G-convergence of sequences of elliptic operators we define the generalized solutions for the problems. The linear parabolic equations as well as the free boundary problems with isotropic elliptic operators defined in a cylinder $\Omega \times (0,\tau) \subset R^{n+1}$, are considered. In Section 4 a regularized parametric optimization problem depending on a parameter $\varepsilon > 0$ is introduced. The method of reqularization assures the existence of an optimal solution $u_\varepsilon \in L^\infty(\Omega)$ for the optimization problems for any $\varepsilon > 0$. The generalized solutions of such problems are defined in the form of the limits with respect to G-convergence for $\varepsilon \downarrow 0$ of the sequence $\{u_\varepsilon I\} \subset L^\infty(\Omega; R^{n^2})$ of matrix functions. The necessary optimality conditions for the optimization problems under consideration are derived.

At the end of this chapter we present the list of references concerning the parametric optimization problems for distributed parameter systems and the related topics.

We refer the reader to [1, 11, 38, 43, 45] for the general results on the modelling and control of systems described by partial differential equations. Problems of control in coefficients for such a systems are considered in [2, 4, 5, 13, 14, 33, 38, 48, 49, 61÷68, 69, 70, 76, 80]. The applications of latter problems in the field of the optimal design are given in [10, 22, 27÷29, 32, 36, 39, 51, 54, 56]. The concept of the generalized solutions of the parametric optimization problems for elliptic systems is used in [3, 10, 22, 28, 51, 54, 57÷60]. We refer the reader to [11, 12, 23, 26, 37, 40 42, 46, 47, 50, 71, 72÷74, 77÷79] for the results concerning G-convergence theory.
Finally let us recall that the identification problems and the inverse problems for the distributed parameter systems are considered in [6÷9, 15÷21, 24, 25, 30, 35, 44, 52, 53, 55, 75].

The outline of this chapter is the following. In section 2 the parametric optimization problem for an abstract evolution equation is introduced. The solution of such a problem is defined and the form of the necessary optimality conditions is derived.

In Section 3 the related results on the G-convergence of sequences of the second order elliptic operators are presented.

Section 4 is devoted to the parametric optimization problems for parabolic equation. The generalized solutions for such problems are introduced.

Finally in Section 5 the parametric optimization problems for free boundary problems are considered.

In this chapter the standard notation in used [45].

2. Parametric optimization problems for abstract evolution equation

In this section the parametric optimization problem for an abstract parabolic equation is considered.

2.1. Abstract parabolic equation

Let V be a reflexive and separable real Banach space and let V' be the dual of V. We shall denote by the same symbol $\|\cdot\|$ both the norm in V and its adjoint norm in V', furthermore $\langle .,. \rangle$ denotes duality pairing between V' and V. We denote by $E(V)$ the class of linear operators $T \in \mathcal{L}(V;V')$ which are symmetric and positive isomorphisms, i.e. such that

$$\langle Tu,v \rangle = \langle Tv,u \rangle, \quad \forall u,v \in V \tag{2.1}$$

$$\alpha_T \|u\|^2 \leq \langle Tu,u \rangle \leq \beta_T \|u\|^2, \quad \alpha_T > 0, \quad \forall u \in V \tag{2.2}$$

We will use in the sequel the concept of the following convergence of sequences in $E(V)$.

<u>Definition 2.1</u>

Let T_k, $T \in E(V)$ be given, $k=1,2,\ldots$. We say that the sequence $\{T_k\}$ is G-convergent to T for $k \to \infty$:

$$T_k \xrightarrow{G} T \tag{2.3}$$

if for any f and g in V':

$$\lim_{k\to\infty} \langle g, T_k^{-1} f \rangle = \langle g, T^{-1} f \rangle \tag{2.4}$$

Remark 2.1

G-convergence is the sequence - convergence for the topology defined on E(V) by the family of semi - norms

$$\{T \longrightarrow < f, T^{-1} g > \mid f, g \in V'\} \tag{2.5}$$

Theorem 2.1

For any sequence $\{T_k\} \subset E(V)$ such that

$$\alpha \|u\|^2 \le < T_k u, u > \le \beta \|u\|^2, \quad \alpha > 0, \quad \forall u \in V \tag{2.6}$$

there exists a subsequence, still denoted $\{T_k\}$ and an element $T \in E(V)$ such that

$$T_k \xrightarrow{G} T \tag{2.7}$$

furthermore

$$\alpha \|u\|^2 \le < Tu, u > \le \beta \|u\|^2, \qquad \forall u \in V \tag{2.8}$$

The proof of Theorem 2.1 is given e.g. in [47, 78]. Let us consider the evolution problem:

$$\frac{dy}{dt} + Ty = f \quad \text{in} \quad L^2(0,\tau;V'), \quad \tau > 0 \tag{2.9}$$

$$y(0) = y^o \quad \text{in} \quad H = [V, V']_{1/2} \tag{2.10}$$

where $f \in L^2(0,\tau;V')$, $y^o \in H$, $T \in E(V)$ are given elements. For the definition of the space $H = [V,V']_{1/2}$ we refer the reader to the book [45]. In particular, in the case of Sobolev spaces: for $V = H_o^1(\Omega)$, $V' = H^{-1}(\Omega)$ we have [45] $H = L^2(\Omega)$.

It is well known [45] that the solution y of the problem (2.9), (2.10) belongs to the space:

$$W(0,\tau) = \{\phi \in L^2(0,\tau;V) \mid \frac{d\phi}{dt} \in L^2(0,\tau;V')\} \tag{2.11}$$

2.2. Parametric optimization problem

In order to define an optimization problem we introduce the function

$$I(.) : E(V) \longrightarrow R \tag{2.12}$$

of the form

$$I(T) = \frac{1}{2} \int_0^\tau \|y(T;t) - z_d(t)\|_H^2 \, dt \tag{2.13}$$

where $z_d(.) \in L^2(0,\tau;H)$ is a given element, $y=y(T;.)$ denotes a unique solution of the evolution problem (2.9), (2.10).

We assume that there is given a convex subset $U \subset E(V)$ which is closed with respect to the G-convergence.

Let us consider the following parametric optimization problem:

Problem (P):

find an element $T \in U$ which minimizes the cost functional (2.13) over the set $U \subset \mathcal{L}(V;V')$.

Theorem 2.2

There exists a solution $T \in U$ to the problem (P). The element $T \in \mathcal{L}(V;V')$ satisfies the following optimality system:

find $(y,p,T) \in W(0,\tau) \times W(0,\tau) \times U$ such that

$$\frac{dy}{dt} + Ty = f \quad \text{in} \quad L^2(0,\tau;V') \qquad y(0) = y^o \quad \text{in} \quad H \tag{2.14}$$

$$-\frac{dp}{dt} + Tp = y - z_d \quad \text{in} \quad L^2(0,\tau;V') \qquad p(\tau) = 0 \quad \text{in} \quad H \tag{2.15}$$

$$\int_0^\tau \langle Ty,p \rangle dt \geq \int_0^\tau \langle Sy,p \rangle dt \qquad \forall S \in U \tag{2.16}$$

Proof

(i) Existence of an optimal solution.

Let $\{T_k\} \subset U$ be a minimizing sequence for the problem (P). From Theorem 2.1 it follows that there exists a subsequence, still denoted $\{T_k\}$ and an element $T \in E(V)$ such that

$$T_k \xrightarrow{G} T \quad \text{with} \quad k \to \infty$$

Since the set $U \subset \mathcal{L}(V;V')$ is closed with respect to G-convergence it follows that $T \in U$.

Denote by $\{y_k\} \subset W(0,\tau)$ the sequence of solutions to the evolution equation

$$\frac{dy_k}{dt} + T_k y_k = f \quad \text{in} \quad L^2(0,\tau;V') \qquad y_k(0) = y^o \quad \text{in} \quad H \tag{2.17}$$

From Theorem 2.3 in [71] it follows that

$$y_k \rightharpoonup y \quad \text{weakly in} \quad W(0,\tau) \quad \text{with} \quad k\to\infty, \tag{2.18}$$

where the element y satisfies the evolution equation (2.9), (2.10).

Since the imbedding $W(0,\tau) \subset L^2(0,\tau;H)$ is continuous [45], by (2.18) and by lower semicontinuity of the norm in space $L^2(0,\tau;H)$ it follows that

$$\liminf_{k\to\infty} I(T_k) \geq I(T) \tag{2.19}$$

hence the element $T \in U$ is an optimal solution of the problem (P).

(ii) <u>Necessary optimality conditions</u>

Let us denote

$$\forall S \in U : dI(T;S) \stackrel{\text{def}}{=} \lim_{\varepsilon\downarrow 0} (I(T+\varepsilon S) - I(T))/\varepsilon \tag{2.20}$$

let $p \in W(0,\tau)$ be the element given by a unique solution of the adjoint evolution equation (2.15).

Simple calculations show that we have

$$dI(T;S) = \int_0^\tau \langle Sy,p \rangle dt \qquad \forall S \in E(V) \tag{2.21}$$

where the element $y \in W(0,\tau)$ is given by a unique solution of the equation (2.9), (2.10).

The form of the optimality system (2.14)-(2.16) can be obtained by a standard argument, taking into account (2.21). ■

3. G-convergence of elliptic operators

This section is concerned with the G-convergence of sequences of the second order elliptic operators.

We introduce the necessary notation. Let $\Omega \subset R^n$ be a given domain with smooth boundary $\partial\Omega$. We denote by

$$T = T(A) \in \mathcal{L}(H_0^1(\Omega);H^{-1}(\Omega)) \tag{3.1}$$

the elliptic operator

$$T = -\sum_{i,j=1}^{n} \frac{\partial}{\partial x_i}(a_{ij} \frac{\partial}{\partial x_j}) \tag{3.2}$$

where $a_{ij}(\cdot)$, $i,j=1,\dots,n$ are real, measurable functions defined on Ω such that:

$$a_{ij}(x) = a_{ji}(x)$$

$$0 < \alpha \leq \sum_{i,j=1}^{n} a_{ij}(x)\xi_i\xi_j|\xi|^{-2} \leq \beta , \quad \forall \xi \in R^n \tag{3.3}$$

$$\text{for a.e.} \quad x \in \Omega$$

The set of all matrix functions $A=[a_{ij}]_{nxn}$ which satisfy (3.3) will be denoted in the sequel by $E_{\alpha,\beta}$. The functions $a_{ij}(\cdot)$ are called the coefficients of the operator $T=T(A)$. Let $A=[a_{ij}]_{nxn}$, $B=[b_{ij}]_{nxn} \in E_{\alpha,\beta}$ be given matrix functions. It can be shown [46] that if

$$\int_\Omega a_{ij}(x) \frac{\partial y}{\partial x_i}(x) \frac{\partial z}{\partial x_j}(x)dx \overset{\text{def}}{=} \int_\Omega \langle A(x).\nabla y(x), \nabla z(x)\rangle_{R^n} dx$$

$$= \int_\Omega \langle B(x).\nabla y(x), \nabla z(x)\rangle_{R^n} dx, \quad \forall y,z \in H_o^1(\Omega) \tag{3.4}$$

then

$$a_{ij}(x) = b_{ij}(x) \quad \text{for a.e.} \quad x \in \Omega, \text{ and for all } i,j=1,\dots,n \tag{3.5}$$

We define the G-convergence of a sequence $\{A_k\} \subset E_{\alpha,\beta}$ to an element $A \in E_{\alpha,\beta}$.

Definition 3.1

Let there be given a sequence $\{A_k\} \subset E_{\alpha,\beta}$, $k=1,2,\dots$. We say that the sequence $\{A_k\}$ is G-convergent in Ω to an element $A_\infty \in E_{\alpha,\beta}$ for $k \to \infty$ and we denote

$$A_k \xrightarrow{G} A \quad \text{in} \quad \Omega \tag{3.6}$$

if

$$\forall f \in H^{-1}(\Omega) : T^{-1}(A_k)f \longrightarrow T^{-1}(A)f \quad \text{weakly in} \quad H_o^1(\Omega) \tag{3.7}$$

It can be shown [74] that the set $E_{\alpha,\beta}$ of matrix functions is sequentially compact with respect to the G-convergence.

Theorem 3.1

Let $\{A_k\} \subset E_{\alpha,\beta}$, $k=1,2,\dots$ be a given sequence. Then there exists a subsequence, still denoted $\{A_k\}$, and an element $A_\infty \in E_{\alpha,\beta}$ such that for $k \to \infty$:

$$A_k \xrightarrow{G} A_\infty \quad \text{in} \quad \Omega \tag{3.8}$$

The proof of Theorem 3.1 is given e.g. in [73].

3.1. G-convergence of isotripic operators

Let us consider the G-convergence for the particular class of the so-called isotropic operators:

$$T(A) = T(uI) = -\mathrm{div}(u\nabla) \tag{3.9}$$

here $A=uI$, I is the identity matrix and $u \in U_{ad} \subset L^{\infty}(\Omega)$ where

$$U_{ad} = \{v \in L^{\infty}(\Omega) \mid 0 < \alpha \leq v(x) \leq \beta \quad \text{for a.e.} \quad x \in \Omega\} \tag{3.10}$$

The following Theorem 3.2 describes G-limits of a sequence of isotropic operators.

Theorem 3.2

Let $\{u_k\} \subset U_{ad}$, $k=1,2,\dots$ be a given sequence. Then there exists a subsequence, still denoted $\{u_k\}$, and elements $u_{\infty}, v_{\infty} \in U_{ad}$, $A \in E_{\alpha,\beta}$ such that for $k \to \infty$:

$$u_k \longrightarrow u_{\infty} \quad \text{weakly - (*) in } L^{\infty}(\Omega), \tag{3.11}$$

$$1/u_k \longrightarrow 1/v_{\infty} \quad \text{weakly - (*) in } L^{\infty}(\Omega), \tag{3.12}$$

$$u_k I \xrightarrow{G} A \quad \text{in } \Omega. \tag{3.13}$$

Let $u_{\infty}(x) = \alpha\theta(x) + \beta(1-\theta(x))$, $0 \leq \theta(x) \leq 1$, $x \in \Omega$. For a.e. $x \in \Omega$ the eigenvalues $\mu_j = \lambda_j(A)(x)$, $j=1,\dots,n$ of the matrix $A(x)$ satisfy the following conditions:

$$\alpha \leq \mu_j \leq \beta, \qquad j=1,\dots,n \tag{3.14}$$

$$\sum_{j=1}^{n} \frac{1}{\mu_j - \alpha} \leq \frac{1}{\mu_-(\theta) - \alpha} + \frac{n-1}{\mu_+(\theta) - \alpha}, \tag{3.15}$$

$$\sum_{j=1}^{n} \frac{1}{\beta - \mu_j} \leq \frac{1}{\beta - \mu_-(\theta)} + \frac{n-1}{\mu_+(\theta) - \alpha}, \tag{3.16}$$

where $\theta = \theta(x)$, $\mu_+(\theta) = \alpha\theta + \beta(1-\theta)$, $\mu_-(\theta) = 1/(\theta/\alpha + (1-\theta)/\beta)$.

The proof of Theorem 3.2 is given in [41, 50, 77]. ∎

It should be noted here that in general the G-limit of a sequence of isotropic operators is not an isotropic operator.

We denote by GU_{ad} the G-closure of the set U_{ad} defined in the following way:

$$GU_{ad} = \{A \in E_{\alpha,\beta} \mid \exists \{u_k\} \subset U_{ad} \text{ such that}$$

$$u_k I \xrightarrow{G} A \text{ in } \Omega\} \tag{3.17}$$

Lemma 3.1 [40, 76]

Let us assume that $\Omega \subset R^2$. Then the set GU_{ad} is given by

$$GU_{ad} = \{A \in E_{\alpha,\beta} \mid \text{for a.e. } x \in \Omega \text{ the eigenvalues} \tag{3.18}$$

$$\lambda_1(x),\ \lambda_2(x) \text{ of the matrix } A(x) \text{ satisfy the condition}$$

$$\alpha \le \frac{\alpha\beta}{\alpha+\beta-\lambda_2(x)} \le \lambda_1(x) \le \lambda_2(x) \le \beta\}$$

The proof of Lemma 3.1 follows from Theorem 3.2

Lemma 3.2

The set (3.18) is convex.
The proof of Lemma 3.2 is given in [71].

3.2. Boundary-value problems-applications of G-convergence

Let there be given a sequence $\{u_k\} \subset U_{ad}$. We assume that for $k \to \infty$

$$u_k I \xrightarrow{G} A \text{ in } \Omega$$

where $A \in GU_{ad}$ is a given element.

We present some results on the convergence of solutions to some boundary-value problems of elliptic and parabolic types. We will consider boundary-value problem of elliptic type such as the obstacle problem, as well as boundary-value problems of parabolic type such as the parabolic equation and the evolution unilateral problem.

3.3. The obstacle problem

Let us consider a sequence of solutions of variational inequality:

Find $y_k \in H_0^1(\ )$, $k=1,2,\ldots,$ such that

$$y_k \in K(\Omega) = \{\phi \in H_0^1(\Omega) \mid \phi(x) \ge 0 \text{ a.e. on } \Omega\}$$

$$\int_\Omega u_k(x)\nabla y_k(x).\nabla(\phi(x)-y_k(x))dx \geq \int_\Omega f(x)(\phi(x)-y_k(x))dx \qquad (3.19)$$

$$\forall \phi \in K(\Omega)$$

where $f \in H^{-1}(\Omega)$ is a given element in $H^{-1}(\Omega)$, $\{u_k\} \subset U_{ad}$ is a sequence such that the condition (3.13) is verified.

By the results of Boccardo and Murat [12] there exists an element $y \in H_o^1(\Omega)$ such that the following convergences take place for $k \to \infty$:

$$y_k \rightharpoonup y \quad \text{weakly in} \quad H_o^1(\Omega) \qquad (3.20)$$

$$u_k \nabla y_k \rightharpoonup A.\nabla y \quad \text{weakly in} \quad L^2(\Omega;R^n)\ , \qquad (3.21)$$

$$u_k \nabla y_k . \nabla y_k \rightharpoonup \langle A.\nabla y\ ,\ \nabla y\rangle_{R^n} \quad \text{in} \quad \mathcal{D}'(\Omega)\ . \qquad (3.22)$$

The element y is given by a unique solution of the following variational inequality:

$$y \in K(\Omega)$$

$$\int_\Omega \langle A(x).\nabla y(x), \nabla(\phi(x)-y(x))\rangle_{R^n} dx \geq \int_\Omega f(x)(\phi(x)-y(x))dx \qquad (3.23)$$

$$\forall \phi \in K(\Omega)$$

3.4. <u>Parabolic equation</u>

Let us consider a sequence of solutions of parabolic initial-boundary-value problem:

$$\frac{\partial y_k}{\partial t}(x,t) - \operatorname{div}(u_k(x)\nabla y_k(x,t)) = f(x,t) \quad \text{in} \quad \Omega \times (0,\tau) \qquad (3.24)$$

$$y_k(x,t) = 0 \quad \text{on} \quad \partial\Omega \times (0,\tau) \qquad (3.25)$$

$$y_k(x,0) = y^o(x) \quad \text{on} \quad \Omega \qquad (3.26)$$

where $f \in L^2(\Omega \times (0,\tau))$, $y^o \in L^2(\Omega)$ are given elements, $\{u_k\} \subset U_{ad}$ is a given sequence such that condition (3.13) is verified.
The solution y_k belongs to the space $W(0,\tau)$, i.e.

$$y_k \in L^2(0,\tau;H_o^1(\Omega))$$

$$\partial y_k/\partial t \in L^2(0,\tau;H^{-1}(\Omega))$$

It can be shown [23, 79] that there exists an element $y \in W(0,\tau)$ such that the following convergences take place for $k \to \infty$:

$$y_k \longrightarrow y \text{ weakly in } L^2(0,\tau;H_o^1(\Omega)) \text{ and strongly in } L^2(\Omega \times (0,\tau)) \tag{3.27}$$

$$\partial y_k/\partial t \longrightarrow \partial y/\partial t \text{ weakly in } L^2(0,\tau;H^{-1}(\Omega)) \tag{3.28}$$

The element y satisfies the following parabolic equation:

$$\frac{\partial y}{\partial t}(x,t) - \operatorname{div}(A(x).\nabla y(x,t)) = f(x,t) \quad \text{in} \quad \Omega \times (0,\tau) \tag{3.29}$$

$$y(x,t) = 0 \qquad \text{on} \quad \partial\Omega \times (0,\tau) \tag{3.30}$$

$$y(x,0) = y^o(x) \qquad \text{on} \quad \Omega \tag{3.31}$$

3.5. Evolution variational inequality

Let us consider a sequence of solutions of the variational inequality:

Find y_k, $k=1,2,...$ such that

$$y_k \in K(Q) = \{\phi \in W(0,\tau) \mid \phi(x,t) \geq 0 \text{ a.e. on } Q = \Omega \times (0,\tau)\}$$

$$\int_\Omega \frac{\partial y_k}{\partial t}(x,t)(\phi(x)-y_k(x,t))dx + \int_\Omega u_k(x)\nabla y_k(x,t).\nabla(\phi(x)-y_k(x,t))dx$$

$$\geq \int_\Omega f(x,t)(\phi(x)-y_k(x,t))dx, \quad \text{for a.e. } t \in (0,T), \ \forall\phi \in K(\Omega)$$

$$y_k(x,0) = y^o(x) \quad \text{in} \quad \Omega \tag{3.32}$$

where $f \in L^2(Q)$, $y^o \in L^2(\Omega)$, $y^o(x) \geq 0$ a.e. on Ω, are given elements, $\{u_k\} \subset U_{ad}$ is a given sequence such that the condition (3.13) is verified.

It can be verified [11] that there exists an element $y \in W(0,\tau)$ such that the following convergences take place for $k \to \infty$:

$$y_k \longrightarrow y \text{ weakly in } W(0,\tau) \text{ and strongly in } L^2(Q) \tag{3.33}$$

The element y satisfies the following variational inequality:

$$y \in K(Q) \tag{3.34}$$

$$\int_\Omega \frac{\partial y}{\partial t}(x,t)(\phi(x)-y(x,t))dx + \int_\Omega \langle A(x).\nabla y(x,t), \nabla(\phi(x)-y(x,t))\rangle_{R^n} dx$$

$$\geq \int_\Omega f(x,t)(\phi(x)-y(x,t))dx \quad \text{for a.e. } t \in (0,T), \quad \forall\phi \in K(\Omega)$$

Remark 3.1

The above examples show that G-convergence of a sequence of elliptic operators implies the weak convergence of the corresponding sequence of solutions of boundary-value problems. It should be noted here that such results can be obtained for evolution problems provided that the elliptic operators do not depend on the variable t. We refer the reader to [11, 71] for further examples.

4. Parametric optimization problems for parabolic equation

This section is concerned with the identification problems for parabolic equations.

We assume for simplicity that $\Omega \subset R^2$. Let $f \in L^2(Q)$, $y^o \in L^2(\Omega)$ be given elements. We denote by $y=y(u;x,t)$, $u \in U_{ad}$, $(x,t) \in Q$, the solution of the parabolic equation:

$$\frac{\partial y}{\partial t}(u;x,t)-\mathrm{div}(u(x)\nabla y(u;x,t)) = f(x,t) \quad \text{in } Q \tag{4.1}$$

$$y(u;x,t) = 0 \quad \text{on } \Sigma \tag{4.2}$$

$$y(u;x,0) = y^o(x) \quad \text{on } \Omega \tag{4.3}$$

Here the set U_{ad} is given by (3.10). We introduce the cost functional

$$J(u) = \frac{1}{2}\int_Q (y(u;x,t) - z_d(x,t))dQ \qquad \forall u \in U_{ad} \tag{4.4}$$

where $z_d \in L^2(Q)$ is a given element.

Let us consider the following optimization problem:

(P) Find an element $u_{opt} \in U_{ad}$ which minimizes the cost functional (4.4) over the set U_{ad}.

It can be verified that the functional (4.4) is differentiable everywhere on an open neighbourhood of the set U_{ad} in $L^\infty(\Omega)$ therefore if there exists an optimal solution of the problem (P) then the solution satisfies the necessary optimality conditions in the form of the following variational inequality:

$$\begin{aligned} & u_{opt} \in U_{ad} \\ & dJ(u_{opt}; u - u_{opt}) \geq 0, \quad \forall u \in U_{ad} \end{aligned} \tag{4.5}$$

Variational inequality (4.5) is equivalent to the following optimality condition

$$u_{opt} \in U_{ad}$$
$$\int_Q (u_{opt}(x)-u(x))\nabla y(u_{opt},x,t).\nabla p(x,t)\,dxdt \geq 0, \ \forall u \in U_{ad} \tag{4.6}$$

where the element $p \in W(0,\tau)$ is given by a solution of the adjoint state equation:

$$-\frac{\partial p}{\partial t}(x,t)-\operatorname{div}(u_{opt}(x)\nabla p(x,t))=y(u_{opt};x,t)-z_d(x,t) \quad \text{in } Q \tag{4.7}$$

$$p(x,t) = 0 \qquad \text{on } \Sigma \tag{4.8}$$

$$p(x,\tau) = 0 \qquad \text{on } \Omega \tag{4.9}$$

Let us recall [48, 67] that in general the existence of an optimal solution of the problem (P) cannot be assured. We present counterexample.

Example 4.1

Let us consider the general case of the coefficient $u=u(x,t)$, $(x,t) \in Q$ We take the set of admissible coefficients of the form:

$$\tilde{U}_{ad} = \{u \in L^\infty(Q) \mid 1-\frac{1}{\sqrt{2}} \leq u(x,t) \leq 1+\frac{1}{\sqrt{2}}, \quad \text{a.e. in } Q\} \tag{4.10}$$

The domain $\Omega \subset R^2$ is given by

$$\Omega = (-2,2) \times (-2,2) \tag{4.11}$$

Denote by $z=z(x)$, $x=(x_1,x_2) \in \Omega$ the function defined as follows

$$z(x) = \begin{cases} (x_1^2+x_2^2-1)^2, & x_1^2+x_2^2 \leq 1 \\ 0 & x_1^2+x_2^2 > 1 \end{cases} \tag{4.12}$$

furthermore denote

$$f(x,t) = -\frac{1}{2}\frac{\partial^2 z}{\partial x_1^2}(x) - \frac{\partial^2 z}{\partial x_2^2}(x), \quad x \in \Omega,\ t \in (0,\tau) \tag{4.13}$$

$$y^o(x) = z(x), \qquad x \in \Omega \tag{4.14}$$

Let us consider the optimization problem:

Problem (P):

Find an element $u_{opt} \in \tilde{U}_{ad}$ which minimizes the functional

$$J(u) = \frac{1}{2} \int_Q (y(u;x,t) - z(x))^2 dxdt \qquad (4.15)$$

over the set (4.10) of admissible coefficients.

Theorem 4.1 [67]

We have

(i) $\inf \{J(u) \mid u \in \tilde{U}_{ad}\} = 0$ (4.16)

(ii) the element $\bar{u} \in \tilde{U}_{ad}$ such that $y(\bar{u};x,t) = z(x)$ for a.e. $(x,t) \in \Omega \times (0,\tau)$

does not exists.

The proof of Theorem 4.1 is given in [67].

4.1. Regularization of parametric optimization problem

We present a method [61, 67, 68] which allows us to assure the existence of an optimal solution to the problem (P).

Let us recall that the norm in Sobolev space $H^s(\Omega)$, $s \in (0,1/2)$ is given by

$$\|u\|_{H^s(\Omega)} = (\omega_s(u))^{1/2} \qquad (4.17)$$

where

$$\omega_s(u) \overset{def}{=} \int_\Omega \int_\Omega (u(x)-u(z))^2/(x-z)^{2s} dxdt + \int_\Omega (u(x))^2 dx$$

For a fixed constants $\varepsilon > 0$, $s \in (0,1/2)$ let us consider, the following optimization problem

Problem (P_ε):

Find an element $u_\varepsilon \in U_{ad} \cap H^s(\Omega)$ which minimizes the cost functional

$$J_\varepsilon(u) = J(u) + \varepsilon\omega_s(u) \qquad (4.18)$$

over the set $U_{ad} \cap H^s(\Omega)$.

Theorem 4.2

For any $\varepsilon > 0$, $s > 0$ there exists an optimal solution $u_\varepsilon \in U_{ad} \cap H^s(\Omega)$

of the problem (P_ε). The element u_ε satisfies the optimality condition:

$$\int_Q (u_\varepsilon(x) - u(x))\nabla y(u_\varepsilon;x,t).\nabla p(x,t)\,dxdt \tag{4.19}$$

$$+ 2\varepsilon \int_\Omega\int_\Omega \{(u_\varepsilon(x)-u_\varepsilon(z))(u(x)-u(z)-u_\varepsilon(x)-u_\varepsilon(z))/(x-2)^{2s}\}dxdz$$

$$+ 2\varepsilon \int_\Omega u_\varepsilon(x)(u(x)-u_\varepsilon(x))dx \geq 0, \qquad \forall u \in U_{ad} \cap H^s(\Omega)$$

here $p=p(x,t)$, $(x,t) \in Q$ denotes the solution of the equation

$$-\frac{\partial p}{\partial t}(x,t)-\mathrm{div}(u_\varepsilon(x)\nabla p(x,t))=y(u_\varepsilon;x,t)-z_d(x,t) \quad \text{in} \quad Q \tag{4.20}$$

$$p(x,t) = 0 \quad \text{on} \quad \Sigma \tag{4.21}$$

$$p(x,\tau) = 0 \quad \text{on} \quad \Omega \tag{4.22}$$

The proof of Theorem 4.2 is omitted here.

■

4.2. Generalized solutions of parametric optimization problem

We define a generalized solution of the problem (P) in the form of an element $A \in E_{\alpha,\beta}$. Let us consider the sequence $\{u_\varepsilon\} \subset U_{ad} \cap H^s(\Omega)$ of solutions of problems (P_ε) for $\varepsilon \downarrow 0$.

Theorem 4.3

Let $\Omega \subset R^n$ be a bounded domain. There exists a subsequence, still denoted $\{u_\varepsilon\}$ and elements $u_o, v_o \in U_{ad}$, $A \in E_{\alpha,\beta}$ such that for $\varepsilon \downarrow 0$:

$$u_\varepsilon \longrightarrow u_o \quad \text{weakly - } (*) \quad \text{in} \quad L^\infty(\Omega), \tag{4.23}$$

$$1/u_\varepsilon \longrightarrow 1/v_o \quad \text{weakly - } (*) \quad \text{in} \quad L^\infty(\Omega), \tag{4.24}$$

$$u_\varepsilon I \xrightarrow{G} A \quad \text{in} \quad \Omega, \tag{4.25}$$

furthermore the eigenvalues $\mu_j=\lambda_j(A)(x)$, $j=1,\ldots,n$ of the matrix $A(x)$ satisfy the conditions (3.14)-(3.16) where the element $\theta=\theta(x)$, $x \in \Omega$, is defined by $\alpha\theta(x)+\beta(1-\theta(x))=u_o(x)$, $x \in \Omega$. The proof of Theorem 4.3 follows from Theorem 3.2.

Definition 4.1

The element $A \in E_{\alpha,\beta}$ which is given by

$$u_\varepsilon I \xrightarrow{G} A \quad \text{in } \Omega \quad \text{for} \quad \varepsilon \downarrow 0 \tag{4.26}$$

is called the generalized solution of the problem (P). ■

Let us introduce the cost functional:

$$I(A) = \frac{1}{2} \int_Q (y(A;x,t) - z_d(x,t))^2 dxdt , \quad \forall A \in E_{\alpha,\beta} \tag{4.27}$$

where $y=y(A,x,t)$, $A \in E_{\alpha,\beta}$, $(x,t) \in Q$, denotes the solution of the parabolic equation:

$$\frac{\partial y}{\partial t}(A;x,t) - \operatorname{div}(A(x).\nabla y(A;x,t)) = f(x,t) \quad \text{in } Q \tag{4.28}$$

$$y(A;x,t) = 0 \qquad \text{on } \Sigma \tag{4.29}$$

$$y(A;x,0) = y^o(x) \qquad \text{on } \Omega \tag{4.30}$$

We denote by $GU_{ad} \subset E$, the G-closure of the set U_{ad} defined by (3.16).

It can be verified that

$$\begin{aligned} &\inf \{J(u) \mid u \in U_{ad}\} = \\ &\inf \{I(uI) \mid u \in U_{ad}\} = \\ &\min \{I(A) \mid A \in GU_{ad}\} \end{aligned} \tag{4.31}$$

Theorem 4.4

There exists an element $A \in GU_{ad}$ which minimizes the cost functional $I(A)$ over the set GU_{ad}. The element A is the generalized solution for the problem (P).

The proof of Theorem 4.4 is omitted here ■

We recall [41, 50, 77] that for isotropic operators the set GU_{ad} is given by (3.14)-(3.16). Let us consider the particular case of $\Omega \subset R^2$. The set GU_{ad} is given by (3.18).

Theorem 4.5

(i) Any generalized solution $A \in GU_{ad}$ of the problem (P) satisfies the necessary optimality condition

$$\int_Q \langle (A(x)-B(x))\, y(A;x,t),\; p(x,t)\rangle_{R^2} dxdt \geq 0, \quad \forall B \in GU_{ad} \tag{4.32}$$

here the element $p=p(x,t)$, $(x,t)\in Q$ denotes the solution of the adjoint equation:

$$-\frac{\partial p}{\partial t}(x,t)-\mathrm{div}(A(x).\nabla p(x,t))=y(A;x,t)-z_d(x,t) \quad \text{in } Q \tag{4.33}$$

$$p(x,t) = 0 \quad \text{on } \Sigma \tag{4.34}$$

$$p(x,\tau) = 0 \quad \text{on } \Omega \tag{4.35}$$

(ii) If there exists an optimal solution $u_o\in U_{ad}$ of the problem (P) then the element u_o satisfies the necessary optimality condition

$$\int_Q u_o(x)\nabla y(u_o;x,t).\nabla p(x,t)\,dxdt \geq \int_Q \langle B(x).\nabla y(u_o;x,t),\nabla p(x,t)\rangle_{R^2}\,dxdt$$

$$\forall B\in GU_{ad} \tag{4.36}$$

where the element $p=p(x,t)$, $(x,t)\in Q$ is given by the solution of (4.20)-(4.22) for $\varepsilon=0$. The proof of Theorem 4.5 is omitted here. ■

5. Parametric optimization problems for variational inequality

This section is concerned with the parametric optimization problems.

5.1. Evolution variational inequality

Let $\Omega\subset R^n$ be a domain with smooth boundary $\partial\Omega$. We denote by $y=y(A;x,t)$, $A\in E_{\alpha,\beta}$, $(x,t)\in Q$, a unique solution of the variational inequality:

$$y(A;.,.)\in K(Q)$$

$$\int_\Omega \{\frac{\partial y}{\partial t}(A;x,t)(\phi(x)-y(A;x,t))+\nabla y(A;x,t).\nabla(\phi(x)-y(A;x,t))\}dx \geq$$

$$\int_\Omega f(x,t)(\phi(x)-y(A;x,t))dx, \quad \forall\phi\in K(\Omega) \text{ for a.e. } t\in(0,T) \tag{5.1}$$

$$y(A;x,0) = y^o(x) \quad \text{on } \Omega, \tag{5.2}$$

here $y^o\in L^2(\Omega)$ is a given element such that $y^o(x)\geq 0$ for a.e. $x\in\Omega$.

For a given element $u\in U_{ad}$ we denote by $y(u;.,.)$ the solution of variational inequality (5.1), (5.2) for $A=uI$.

Let us consider the following optimization problem

Problem (P)

Find an element $u \in U_{ad}$ which minimizes the cost functional (4.4) over the set U_{ad}.

We introduce the regularized problem (P_ε) of the form:

Problem (P_ε)

Find an element $u_\varepsilon \in U_{ad} \cap H^s(\Omega)$ which minimizes the cost functional (4.19) over the set $U_{ad} \cap H^s(\Omega)$.

Here $\varepsilon > 0$, $s \in (0,1/2)$ are given constants.

Theorem 5.1

(i) For any fixed $\varepsilon > 0$, $s \in (0,1/2)$ there exists an optimal solution $u_\varepsilon \in U_{ad} \cap H^s(\Omega)$ of the problem (P_ε)

(ii) There exists a subsequence, still denoted $\{u_\varepsilon\}$ and an element $A \in GU_{ad}$ such that for $\varepsilon \downarrow 0$

$$u\, I \xrightarrow{G} A \quad \text{in } \Omega \tag{5.3}$$

The element A is a generalized solution of the problem (P).

The proof of Theorem 5.1 is omitted here.

5.2. Obstacle problem

We recall the results on conical differentiability of solutions of the obstacle problem with respect to the coefficients of the elliptic operator.

We denote by $y_\mu \in H_0^1(\Omega)$, $\mu > 0$, the solution of the variational inequality:

$$y_\mu \in K(\Omega)$$

$$\int_\Omega \langle (A(x) + \mu B(x))\nabla y_\mu(x), \nabla(\phi(x) - y_\mu(x))\rangle_{R^n} dx \geq \int_\Omega f(x)(\phi(x) - y_\mu(x))dx, \quad \forall \phi \in K(\Omega) \tag{5.4}$$

here $A \in E_{\alpha,\beta}$ is a given element and we assume that the matrix function B is such that for $\mu > 0$, μ small enough $(A + \mu B) \in E_{\alpha,\beta}$.

We present the following result [70, 71] on the right-differentiability of the element $y_\mu \in H_o^1(\Omega)$ with respect to μ at $\mu=0$.

Theorem 5.2 [70]

For $\mu > 0$, μ small enough

$$y_\mu = y_o + \mu q + o(\mu) \quad \text{in} \quad H_o^1(\Omega) \tag{5.5}$$

where $\|o(\mu)\|_{H_o^1(\Omega)}/\mu \to 0$ with $\mu \downarrow 0$.

The element $q \in H_o^1(\Omega)$ is given by a unique solution of the following variational inequality:

$$q \in S$$

$$\int_\Omega \langle A(x)\nabla.q(x), \nabla(\phi(x)-q(x))\rangle_{R^n} dx \geq - \int_\Omega \langle B(x).\nabla y_o(x), \nabla(\phi(x)-q(x))\rangle_{R^n} dx, \quad \forall \phi \in S \tag{5.6}$$

here we denote by $S \subset H_o^1(\Omega)$ the cone of the form:

$$S = \{\phi \in H_o^1(\Omega) \mid \phi(x) \geq 0 \quad \text{q.e. on} \quad Z(y_o), \quad \int_\Omega \nabla y_o(x).\nabla\phi(x)dx = \int_\Omega f(x)\phi(x)dx\} \tag{5.7}$$

where

$$Z(y_o) = \{x \in \Omega \mid y_o(x) = 0\} \tag{5.8}$$

The proof of Theorem 5.2 is based on the results of F. Mignot and of the author [70, 71].

We assume for simplicity that $\Omega \subset R^2$. Let us consider a parametric optimization problem for the variational inequality:

$$y(u;.) \in K(\Omega)$$

$$\int_\Omega u(x)\nabla y(u;x).\nabla(\phi(x)-y(u;x))dx \geq \int_\Omega f(x)(\phi(x)-y(u;x))dx \quad \forall \phi \in K(\Omega) \tag{5.9}$$

here $u \in U_{ad}$, $f \in H^{-1}(\Omega)$.

Let $z_d \in L^2(\Omega)$ be a given element, we define the cost functional of the form:

$$J(u) = \frac{1}{2}\int_\Omega (y(u;x) - z_d(x))^2 dx \tag{5.10}$$

We introduce the following optimization problem.

Problem (P):

Find an element $u \in U_{ad}$ which minimizes the cost functional (5.10) over the set U_{ad}.

We need the following notation:

- for any element $A \in E_{\alpha,\beta}$ we denote by $S_A \subset H_0^1(\Omega)$ the cone:

$$S_A = \{\phi \in H_0^1(\Omega) \mid \phi(x) \geq 0 \quad \text{q.e. on} \quad Z(y(A)), \quad \int_\Omega \langle A(x).\nabla y(A;x), \nabla\phi(x)\rangle_{R^2} dx = \int_\Omega f(x)\phi(x)dx\} \tag{5.11}$$

- for any elements $A, B \in E_{\alpha,\beta}$ we denote by $q=q(A-B) \in H_0^1(\Omega)$ the solution of the variational inequality:

$$q \in S_A : \int_\Omega \langle A(x).\nabla q(x), \nabla(\phi(x)-q(x))\rangle_{R^2} dx \geq \int_\Omega f(x)(\phi(x)-q(x))dx, \quad \forall\phi \in S_A \tag{5.12}$$

furthermore we denote by $\Omega(A-B)$ the coincidence set

$$\Omega(A-B) = \{q(A-B;x) = 0\} \tag{5.13}$$

and by $V=V(A-B) \subset H_0^1(\Omega)$ the linear subspace:

$$V = \{\phi \in H_0^1(\Omega) \mid \phi(x)=0 \quad \text{q.e. on} \quad \Omega(A-B), \quad \int_\Omega \langle A(x).\nabla y(A;x), \nabla\phi(x)\rangle_{R^2} dx = \int_\Omega f(x)\phi(x)\} \tag{5.14}$$

here q.e. means everywhere with possible exception of a set of capacity zero [70]. The element $y(A;.) \in H_0^1(\Omega)$ is a unique solution of the variational inequality (3.23).

Theorem 5.3

There exists a generalized optimal solution $A \in GU_{ad}$ of the problem (P) which satisfies the necessary optimality conditions:

$$\int_\Omega \langle B(x).\nabla y(A;x), \nabla p(B;x)\rangle_{R^2} \geq 0, \quad \forall B \in GU_{ad}-\{A\} \tag{5.15}$$

where the element $p(B;.) \in H_0^1(\Omega)$ is a unique solution of the variational equation:

$$p(B) \in V(B)\ ,\quad B \in GU_{ad} - \{A\}$$

$$\int_\Omega \langle A(x) . \nabla p(B;x), \nabla\phi(x)\rangle_{R^2} dx = \int_\Omega (y(A;x) - z_d(x))\phi(x)\,dx \qquad (5.16)$$

$$\forall \phi \in V(B)$$

The proof of Theorem 5.3 is given in [70].

6. Concluding remarks

In this chapter the notion of the generalized solution for the parametric optimization problems for evolution equations is introduced. The form of the generalized solutions in the case of isotropic elliptic operators is derived. The results presented here are based on the notion of G-convergence of sequences of elliptic operators. Necessary optimality conditions for the nonconvex optimization problems are presented. The optimization problems considered in this chapter are related to the approximate identification problems of distributed parameter systems.

References

[1] N.U. Ahmed, K.L. Teo: Optimal Control of Distributed Parameter System, Elsevier North Holland, New York, 1981.

[2] N.U. Ahmed: Necessary Conditions of Optimality for a Class of Second-Order Hyperbolic Systems with Spatially Dependent Controls in the Coefficients, JOTA 38, (1982), 423-446.

[3] J.L. Armand: Non homogeneity and anisotropy in structural design, in: Optimization Methods in Structural Design, H. Eschenauer and N. Olhoff Eds., Wissenschaftsverlag, Mannheim, 1983, 256-263.

[4] J.M. Ball, J.E. Marsden, M. Slemrod: Controllability for distributed bilinear systems, SIAM J. Control and Optimization, 20, (1982), 575-597.

[5] J.M. Ball, M. Slemrod: Nonharmonic Fourier Series and the Stabilization of Distributed Semi-Linear Control Systems, Comm. on Pure and Appl. Math. 32, (1979), 555-587.

[6] A. Bamberger, G. Chavent, P. Lailly: About the stability of the

inverse problem in 1-D wave equations - application to the interpretation of seismic profiles, Appl. Math. Optim. 5 (1979), 1-47.

[7] H.T. Banks: A Survey of Some Problems and Recent Results for Parameter Estimation and Optimal Control in Delay and Distributed Parameter Systems, Proc. Conf. on Volterra and Functional Differential Equations, Blacksburg, June, 1981, Marcel Dekker, Inc., New York, 1982, 3-24.

[8] H.T. Banks, P.L. Daniel: Estimation of Variable Coefficients in Parabolic Distributed Systems, LCDS Report # 82-22, Brown University, September, 1982.

[9] H.T. Banks, K. Kunisch: An Approximation Theory for Nonlinear Partial Differential Equations with Applications to Identification and Control, SIAM Journal on Control, 20, 1982, 815-848.

[10] M.P. Bendsøe: Optimization of plates, Thesis, Mathematical Institute, The Technical University of Denmark, Lyngby, 1983.

[11] A. Bensoussan, J.L. Lions, G. Papanicolau: Asymptotic Analysis for Periodic Structures, North-Holland, Amsterdam, 1978.

[12] L. Boccardo, F. Murat: Nouveaux resultats de convergence dans des problems unilateraux, Nonlinear Partial Differential Equations and Their Applications, College de France Seminar, Vol. 2, eds., H. Brezis and J.L. Lions, Research Notes in Mathematics, London, Pitman, 1982.

[13] A. Bogobowicz, J. Sokołowski: Optimal water quality control by treating the effluent at the pollution source and by flow regulation, IAHS-AISH Publ. No. 129, Oxford, (1980), 139-143.

[14] A. Bogobowicz, J. Sokołowski: Modelling and control of water quality in a river section, System modelling and optimization, ed. P. Thoft-Christensen, Lectures Notes in Control and Information Sciences, Vol. 59, 1984.

[15] G. Chavent: Analyse Fonctionalle et Identification de Coefficients Repartis dans les Equations aux Derivees Partielles, These d'Etat, Paris, 1971.

[16] G. Chavent: Identification of distributed parameters, Proc. of the 3rd IFAC Symposium on Identification, Hague, 1973.

[17] G. Chavent, P. Lailly, A. Bamberger: Une application de la theorie du controle a un probleme inverse de sismique, Report Institute Francais du Petrole, 1976.

[18] G. Chavent, M. Dupuy, P. Lemonnier: History Mathing by Use of Optimal Control Theory, Society Petroleum Engineers Journal 15 (1), (1975), 74-86.

[19] G. Chavent: Identification of Functional Parameters in Partial Differential Equation, in: Identification of Parameters in Distributed Systems, Eds., R.E. Goodson and M. Polis , American Society of Mech. Engrs., 1974.

[20] G. Chavent: On the Identification of Distributed Parameter Systems, Proc. 5th IFAC Symposium on Identification and System Parameter Estimation, Darmstadt, Federal Republic of Germany, 1979.

[21] G. Chavent: Local stability of the output least sequare parameter estimation technique, INRIA, Rapport de Recherche, No. 136, Rocquencourt, 1982.

[22] Cheng Keng Tung: Optimal design of solid elastic plates, DCAMM S-Report, No.17, The Technical University of Denmark, Lyngby, 1980.

[23] F. Colombini, S. Spagnolo: Sur la convergence de solutions d'equations paraboliques , J. Math. Pures Appl., Vol. 56, 1977, 263-306.

[24] F. Colonius, K. Kunisch: Stability for parameter estimation in two point boundary value problems, Technical Report, No.50, 1984, Institut fur Mathematik, Technische Universitat Graz.

[25] E. Deuflhard, E. Hairer (eds.): Numerical Treatment of Inverse Problems in Differential and Integral Equations, Birkhauser, Stuttgart, 1983.

[26] E. De Giorgi, S. Spagnolo: Sulla convergenza degli integrali dell'energia per operatori ellittici del secondo ordine, Boll. U.M.I. 8 (1973), 137-158.

[27] E.J. Haug, J.S. Arora: Applied Optimal Design, John Wiley & Sons, New York, 1979.

[28] E.J. Haug, J. Cea, eds.: Optimization of Distributed Parameter

Structures, Sijthoff & Noordhoff, Alphen aan den Rijn, Netherlands, 1981.

[29] E.J. Haug, K.K. Choi: Systematic Occurence of Repeated Eigenvalues in Structural Optimization, JOTA 38 (2)(1982), 251-274.

[30] K. Ito: The Application of Legendre-Tau Approximation to Parameter Identification for Delay and Partial Differential Equations, Proc. 22nd IEEE CDC, San Antonio, 1983, 33-37.

[31] K.H. Hoffmann, J. Sprekels: On the identification of heat conductivity and latent heat in a one-phase Stefan problem, Preprint No. 50, Universität Augsburg, Mathematisches Institut, 1985.

[32] C. Jouron: Sur un probleme d'optimisation ou la contrainte porte sur la frequence fondamentale, R.A.I.R.O. Analyse Numerique 12, (4), (1978), 349-374.

[33] Jaffre Jerome: Analyse numerique de deux problemes de domaine optimale et de controle ponctuel, Thesis, l'Universite, Paris 6, 1975.

[34] C.S. Kubrusly: Distributed Parameter System Identification: A Survey, Int. Journal of Control, Vol. 26, 1977, 509-535.

[35] S. Kitamura, S. Nakagiri: Identifiability of Spatially-varying and Constant Parameters in Distributed Systems of Parabolic Type, SIAM Journal on Control, Vol. 15, 1977, 785-802.

[36] R.V. Kohn, G. Strang: Optimal design and relaxation of variational problems, preprint.

[37] R.V. Kohn, M. Vogelius: A new model for thin plates with rapidly varying thickness, to appear.

[38] K.A. Lurie: Applied Optimal Control of Distributed Systems, Plenum Press, New York, 1984.

[39] K.A. Lurie, A.V. Cherkaev, A.V. Fedorov: Regularization of Optimal Design Problems for Bars and Plates, Part 1 and 2, JOTA 37, (4),(1982),499-544.

[40] K.A. Lurie, A.V. Cherkaev: G-Closure of a Set of Anisotropic Conducting Media in Two Dimensions, Soviet Physics Doklady, 26, (7),(1981), 657-659.

[41] K.A. Lurie, A.V. Cherkaev: Exact estimates of conductivity of a binary mixture of isotropic compounds , Preprint 894, A.F. Ioffe Physical Technical Institute, Leningrad, 1984.

[42] K.A. Lurie, A.V. Cherkaev: The problem of formation of an optimal isotropic multicomponent composite , Preprint 895, A.F. Ioffe Physical Technical Institute, Leningrad, 1984.

[43] J.L. Lions: Controle optimal de systemes gouvernes par des equations aux derivees partielles, Dunod, Paris, 1968.

[44] J.L. Lions: Some Aspects of Modelling Problems in Distributed Parameter Systems, Lectures Notes in Control and Information Sciences, Vol. 1, Springer Veralg, New York, 1978, 11-41.

[45] J.L. Lions, E. Magenes: Problemes aux limites non homogenes et applications, Vol. 1, Dunod, Paris, 1968.

[46] F. Murat: Private communication, 1982.

[47] F. Murat: H-convergence, Seminarie d'analyse fonctionnelle et numerique de l'Universite d'Alger 1977-78, Lectures Notes.

[48] F. Murat: Contre - examples pour divers problemes on le controle intervient dans les coefficients, Annali di Mat. Pura et Appl. ser. 4, 112-113 (1979), 49-68.

[49] F. Murat: Control in Coefficients , in: Encyclopedia of Systems and Control, Pergamon Press, 1983.

[50] F. Murat, L. Tartar: Calcul des variations et homogeneisation , in: Collection de la Direction des Etudes et Recherches d'Electricite de France, Eyrolles, Paris, 1984.

[51] A. Myśliński, J. Sokołowski: Non-differentiable Optimization Problems for Elliptic System, SIAM J. on Control and Optimization, Vol. 23, No. 4, July 1985, 632-648.

[52] K.A. Murphy: A Spline-Based Approximation Method for Inverse Problems for a Hyperbolic System Including Boundary Parameters, Proc. 22nd IEEE CDC, San Antonio, 1983, 46-49.

[53] S. Nakagiri: Identifiability of Linear Systems in Hilbert Spaces, SIAM Journal on Control, Vol. 21, 1983, 501-530.

[54] N. Olhoff, K.A. Lurie, A.V. Cherkaev, A.V. Fedorov: Sliding regimes and anisotropy in optimal design of vibrating axisymmetric plates, Int. J. Solids Structures 17 (1981), No. 10, 931-948.

[55] A. Pierce: Unique Identification of Eigenvalues and Coefficients in a Parabolic Problem, SIAM Journal on Control, Vol. 17, 1979, 494-499.

[56] O. Pironneau: Optimal Shape Design for Elliptic Systems, Springer Verlag, New York, 1983.

[57] U.E. Raitum : The Extensions of Extremal Problems Connected with a Linear Elliptic Equation, Soviet Math. Dokl. 19, (16),(1978), 1342-1345.

[58] U.E. Raitum : On Optimal Control Problems for Linear Elliptic Equations, Soviet Math. Dokl. 20, (1), (1979), 129-132.

[59] U.E. Raitum : Sufficient for the Sets of Solutions of Linear Elliptic Equations to Be Weakly Sequantially Closed, Latyian Mathematical Journal 24 (1980), 142-155, (in Russian).

[60] U. E. Raitum: Extremal properties of layered structures, Math. Operations forsch. Statist., Ser. Optimization, 12 (4), (1981), 563-573, (in Russian).

[61] J. Sokołowski: On parametric optimal control for weak solutions of abstract linear parabolic equations, Control and Cybernetics, 4 (3-4),(1975), 59-84.

[62] J. Sokołowski: On parametric optimal control for a class of linear and quasilinear equations of parabolic type, Control and Cybernetics, 4 (1), (1975), 19-38.

[63] J. Sokołowski: On parametric optimal control for partial differentail equations of parabolic type, Lectures Notes in Computer Sciences, Vol. 41, (1976), Springer-Verlag, New York, 623-633.

[64] J. Sokołowski: Parametric optimization problem for abstract parabolic equation with variable operator domain, IRIA-Laboria, Rapport de Recherche, No. 183, 1976.

[65] J. Sokołowski: Remarks on existence of solutions for parametric optimization problems for partial differential equations of parabolic type, Control and Cybernetic 7, (2),(1978), 47-61.

[66] J. Sokołowski: Control in Coefficients for Parabolic Equations, Lectures Notes in Control and Information Sciences, Vol. 22, Springer-Verlag, New York, 1980, 449-454.

[67] J. Sokołowski: Optimal control in coefficients for weak variational problems in Hilbert space, Appl. Math. Opt. 7 (1981), 283-293.

[68] J. Sokołowski: Control in Coefficients for PDE, Abh. Akad. Wiss. DDR, Jahrgang 1981, No. 2N, Akademia Verlag, Berlin, 287-295.

[69] J. Sokołowski, T. Matsumura, Y. Sakawa: Numerical Solution of a Nonlinear Two-Point Boundary-Value Problem by an Optimization Technique, Control and Cybernetics, 11, (1-2),(1982), 41-56.

[70] J. Sokołowski: Optimal Control in Coefficients of Boundary-Value Problems with Unilateral Constraints, Bulletin de l'Academie Polonaise des Sciences, Serie des Sciences Techniques, 31 , (1-2), (1983), 71-81.

[71] J. Sokołowski: G-convergence and control in coefficients for elliptic boundary-values problems, in: Constructive aspects of optimization, K. Malanowski, K. Mizukami, Polish Scientific Publisher, Warsaw, 1985, 98-120.

[72] S. Spagnolo: Sul limite delle soluzioni di problemi di Cauchy relativi all'equazione del calore, Ann. Scu. Norm. Pisa, 21, (1967), 657-699.

[73] S. Spagnolo: Sulla convergenza di soluzioni di equazioni paraboliche ed ellittiche, Ann. Scu. Norm. Pisa, 22, (1968), 577-597.

[74] S. Spagnolo: Convergence in Energy for Elliptic Operators, Proc. 3rd Symp. Numer. Solut. Part. Diff. Equations, Hubbard, ed., Academic Press, 1976, 469-498.

[75] T. Suzuki: Uniqueness and Nonuniqueness in an Inverse Problem for the Parabolic Equation, Journal of Differential Equations, Vol. 47, 1983, 296-316.

[76] L. Tartar: Problemes de Controle des Coefficents dans les Equations aux Derivees Partielles, Lectures Notes in Computer Sciences, Vol. 41, Springer-Verlag, New York, 1976, 420-426.

[77] L. Tartar: Estimation fines de coefficients homogeneises, to

appear in: Colloque en l'honneur d'E. DE GIORGI, P. Kree ed., Research Notes in Mathematics, Pitman.

[78] V.V. Zhikov, S.M. Kozlov, Kha Tieng Ngoan, O,A. Oleinik: Homogenization and G-convergence of Differential Operators, Uspekhi, Mathematicheskikh Nauk 34, (5), (1979), 65-133.

[79] V.V. Zhikov, S.M. Kozlov, O.A. Oleinik: On G-convergence of the parabolic operators, Uspekhi Mathematicheskikh Nauk 36, (1), (1981), 11-58.

[80] T. Zolezzi: Necessary conditions for optimal controls of elliptic or parabolic problems, SIAM J. Control 10, (1972), 594-607.

Chapter 4

FINITE ELEMENT APPROXIMATION OF AN OPTIMAL DESIGN PROBLEM FOR FREE VIBRATING PLATES

Andrzej Myśliński

1. Introduction

In this paper we shall be concerned with an elastic free vibrating plates. The free vibration of a plate are described by a linear elliptic eigenvalue problem [6, 16]. The smallest eigenvalue of this problem depends on the distribution of the thickness of the plate. It is associated with the square of the fundamental frequency of the plate free vibration [6, 8].

This paper is concerned with an optimal design problem of free vibrating plates. Such problems have numerous applications in different branches of engineering [3, 8, 9, 23]. The goal of the optimization problem considered here is to find such a distribution of the thickness of the plate which maximizes the smallest eigenvalue and satisfies the following constraints: the volume of the plate is constant and the thickness of the plate is bounded. The thickness of the plate is a design variable.

Theoretical aspects of the optimization problems of free vibrating plates were studied by many authors [3, 9, 10, 13, 19, 20, 22, 23, 25, 26]. The existence of solutions and the methods of regularization of this optimization problem were investigated in [9, 13, 19, 20]. The necessary optimality conditions were formulated in [3, 9, 10, 19, 20, 22, 25, 26]. In [25] a descent method was employed to obtain numerical solution of this optimization problem under the assumption that the smallest eigenvalue is simple. In [10, 19, 20, 26] this problem was investigated without this assumption of the uniqueness of the smallest eigenvalue. Moreover in [19, 20] numerical results confirming the multiplicity of the smallest eigenvalues were provided. Note that in the case of multiplicity the optimization problem becomes nondifferentiable.

In this paper we shall concentrate on numerical aspects of this optimization problem. The approximation of the optimization problem of free vibrating plates was consider only in [13] where the convergence of approximation was shown under an assumption that the distribution of the thickness of the plate is a function belonging to the Sobolev space $W^{1,p}$, $p>2$. This space is defined by (1.1). In this paper we employ finite element method as the approximation method of this problem. We shall show the convergence of the proposed approximation for the thickness of the plate belonging to the less regular space $W^{1,2}$. To solve discretized optimization problem without the assumption that the smallest eigenvalue is unique we have to employ a non-smooth optimization method. We use the Lemarchal's method [12] combined with a shifted penalty function method [7]. We also present some numerical results.

We shall use the following notation:

$(\cdot,\cdot)_{R^n}$ is the inner product in R^n

Ω is the open subset in R^n

$W^{m,p}(\Omega)$ is the Sobolev space defined by [2, 11]:

$$W^{m,p}(\Omega) = \{v \in L^p(\Omega) \; : \; D^\alpha v \in L^p(\Omega), |\alpha| \leq m\} \qquad (1.1)$$

where $m \geq 0$, $p \geq 1$ and

$$\alpha = (\alpha_1, \alpha_2, \ldots, \alpha_n)$$

$$\alpha_i \geq 0, \; i=1,\ldots,n, \text{ are integers}$$

$$|\alpha| = \alpha_1 + \alpha_2 + \ldots + \alpha_n$$

$$D^\alpha v = \partial^{|\alpha|} v \; / \; \partial x_1^{\alpha_1} \partial x_2^{\alpha_2} \ldots \partial x_n^{\alpha_n}$$

$L^p(\Omega)$, $1 \leq p < \infty$, is the space of measurable functions that are p-integrable [1]; $H^m(\Omega)=W^{m,2}(\Omega)$ is the Sobolev space with the norm $\|\cdot\|_{H^m(\Omega)}$

V is a linear subspace of the space $H^2(\Omega)$

U is a linear subspace of the space $H^1(\Omega)$

$x = (x_1, x_2)$

$n = (n_1, n_2) = (n_{x_1}, n_{x_2})$ is the outward normal vector to the boundary Γ of the set Ω

$s = (-n_2, n_1)$ is the tangent vector to the boundary Γ of the set Ω

$\frac{\partial y}{\partial n}$ is the outward normal derivative of a function y on Γ

$y_{ij} = \partial^2 y / \partial x_i \partial x_j \qquad i,j=1,2$

$\Delta y = y_{11} + y_{22}$.

2. Problem Formulation

Let be given an elastic vibrating plate (Fig.1). The plate occupies domain $\Omega \subset R^2$ in the plane $x_1 \, x_2$. The domain Ω is bounded, connected and has Lipschitz continuous boundary Γ. The boundary Γ is divided into three parts: Γ_1, Γ_2 and Γ_3. Along Γ_1 the plate is clamped, along Γ_2 the plate is simply supported and along Γ_3 the plate is free.

Let us denote by $u=u(x_1,x_2)$ the variable thickness of the plate and by $y=y(u)=y(u,x_1,x_2)$ the displacement of vibrating plate depending on u [3, 6]. u is our design variable.

The free vibration of an elastic plate are described by linear elliptic eigenvalue problem [6, 16]. This problem consists in finding pairs (λ, y), depending on u such that $\lambda \in R$, $y \neq 0$ satisfy:

$$A_u y = \lambda B_u y \tag{2.1}$$

λ and y denote respectively an eigenvalue and an eigenvector of operator A_u with respect to operator B_u. Since the eigenvalue λ depends on u we shall write it as $\lambda(u)$. The elliptic operator A_u is given by

$$A_u y = (u^3 y_{11})_{11} + (u^3 y_{22})_{22} + \nu(u^3 y_{11})_{22} + \nu(u^3 y_{22})_{11} + 2(1-\nu)(u^3 y_{12})_{12} \tag{2.2}$$

where ν is a Poisson's ratio: ν=const, $0 < \nu < 0.5$.
The operator B_u is given by

$$B_u y = u \, y \tag{2.3}$$

The boundary conditions are given by

$$y = 0 \qquad \frac{\partial y}{\partial n} = 0 \tag{2.4}$$

along the clamped edge Γ_1 and by

$$y = 0 \qquad M(y) = 0 \tag{2.5}$$

along the simply supported edge Γ_2, as well as by

$$M(y) = 0 \qquad T(y) = 0 \tag{2.6}$$

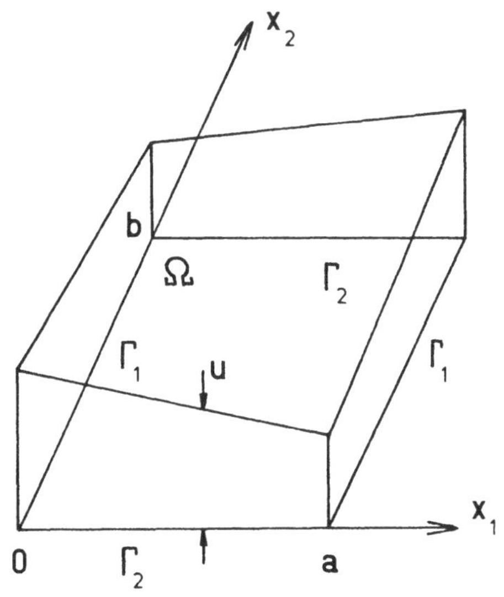

Fig.1. An elastic vibrating plate.

along the free edge Γ_3, where $M(y)$ and $T(y)$ are given by [3, 6, 16]:

$$M(y) = u^3 [\nu \Delta y + (1-\nu)(y_{11} n_1^2 + 2y_{12} n_1 n_2 + y_{22} n_2^2] \tag{2.7}$$

$$T(y) = -\frac{\partial}{\partial n}(u^3 \Delta y) + (1-\nu)\frac{\partial}{\partial s}[u^3 (y_{11} n_1 n_2 - y_{12}(n_1^2 - n_2^2) - y_{22} n_1 n_2)] \tag{2.8}$$

We shall consider weak solutions of the problem (2.1)-(2.8). To this end let us denote:

$$V = \{v \in H^2(\Omega) : v|_\Gamma = 0, \frac{\partial v}{\partial n}\Big|_{\Gamma_1} = 0\} \tag{2.9a}$$

$$U = \{u \in H^1(\Omega) : u \text{ satisfies (2.9c)}\} \tag{2.9b}$$

$$u_{min} \leq u \leq u_{max} \tag{2.9c}$$

where $0 < u_{min} < u_{max}$ are given constants.

We denote by $a_u(y,\phi) : V \times V \longrightarrow R$ and by $b_u(y,\phi) : L^2(\Omega) \times L^2(\Omega) \to R$ the bilinear forms, depending on u, generated by operators (2.2) and (2.3) respectively. These forms are given by [6]:

$$a_u(y,\phi) = \int_\Omega u^3 [y_{11}\phi_{11} + y_{22}\phi_{22} + \nu y_{11}\phi_{22} + \nu y_{22}\phi_{11} + 2(1-\nu) y_{12}\phi_{12}] dx \tag{2.10}$$

$$b_u(y,\phi) = \int_\Omega u y \phi dx \tag{2.11}$$

<u>Lemma 2.1</u>: <u>The bilinear form</u> (2.10), <u>for all</u> $u \in U$, <u>is symmetric</u>, <u>continuous and satisfies the condition: for all</u> $y \in V$ <u>there exists</u> $\alpha > 0$ <u>such that</u>:

$$a_u(y,y) \geq \alpha \, ||y||_{H^2(\Omega)} \tag{2.12}$$

<u>Proof</u>: is given in [18].

Using Green's formula [1, 2] we can write the equation (2.1) with the boundary conditions (2.4)-(2.6) in an equivalent weak form [6, 16]: find pairs $(\lambda, y) \in R \times V \backslash \{0\}$ satisfying:

$$a_u(y,\phi) = \lambda \, b_u(y,\phi) \qquad \forall \phi \in V \tag{2.13}$$

From now on we shall describe the free vibration of the plate by (2.13).

Let us recall that from Riesz-Fredholm theorem [1] and Lemma 2.1 it follows that for any fixed $u \in U$ there exist the sequence $\{\lambda_k\}$ of positive eigenvalues and the sequence $\{y_k\} \subset L^2(\Omega)$ of the corresponding eigenfunctions such that:

(a) for all $k \geq 1$

$$a_u(y_k,\phi) = \lambda_k b_u(y_k,\phi) \quad \text{for all} \quad \phi \in V$$

(b) $0 < \lambda_1 \leq \lambda_2 \leq \dots \leq \lambda_k \dots$

and $\lambda_k \longrightarrow \infty$ for $k \longrightarrow \infty$ (2.14)

(c) $a_u(y_i,y_k) = b_u(y_i,y_k) = 0 \qquad i \neq k$

(d) $a_u(y_i,y_i) = \lambda_i \qquad b_u(y_i,y_i) = 1$

The sequence $\{y_k\}$ constitutes a B_u orthonormal basis in $L^2(\Omega)$. From now on we shall denote the smallest eigenvalue $\lambda(u)$ satisfying (2.13) by $\lambda^*(u)$.
The smallest eigenvalue $\lambda_1 = \lambda^*(u)$ of the problem (2.13) can be also defined by [11]:

$$\begin{aligned} \lambda^*(u) &= \inf\{a_u(y,y) \mid y \in V,\ b_u(y,y) = 1\} \\ &= \inf\{a_u(y,y)/b_u(y,y) \mid y \in V\} \end{aligned} \tag{2.15}$$

We shall also use the following characterization of $\lambda^*(u)$:

<u>Lemma 2.2</u>: <u>There exists a weakly compact set</u> $W \subset V$ <u>such that for every element</u> $u \in U$ <u>the smallest eigenvalue</u> $\lambda^*(u)$ <u>is given by</u>:

$$\lambda^*(u) = \inf\{a_u(y,y) \mid y \in W\} \tag{2.16}$$

<u>Proof</u>: is given in [20].

Let us denote by:

$$I(u) = \lambda^*(u) \tag{2.17}$$

$$R(u) = -0.5\,\varepsilon \|u\|^2_{H^1(\Omega)} \tag{2.18}$$

where $\varepsilon > 0$ is a regularization parameter, and introduce the set of admissible parameters:

$$U_{ad} = \{u \in H^1(\Omega) : u_{min} \leq u \leq u_{max}, \int_\Omega u dx = c\} \tag{2.19}$$

where $0 < u_{min} < u_{max}$, $c > 0$ are given constants such that the set of admissible parameters U_{ad} is nonempty.

We shall consider the following optimization problem:

$$(P) \quad \left| \begin{array}{l} \text{maximize the cost functional} \\ J(u) = I(u) + R(u) \\ \text{subject to } u \in U_{ad} \text{ and the state equation (2.13).} \end{array} \right. \tag{2.20}$$

In next section we shall use the following auxiliary lemma:

<u>Lemma 2.3</u>: <u>The mappings</u>:

$$L^2(\Omega) \supset U_{ad} \ni u \longrightarrow \lambda^*(u) \in R$$

$$L^2(\Omega) \supset U_{ad} \ni u \longrightarrow y(u) \in V$$

<u>where</u> $y(u) \in V$ <u>satisfies</u> (2.14)(c), (d) <u>are continuous</u> .

<u>Proof</u>: is given in [19].

From (2.17)-(2.19) and Lemma 2.3 follows:

<u>Lemma 2.4</u>: <u>There exists an optimal solution</u> $\hat{u} \in U_{ad}$ <u>of problem</u> (P).

<u>Proof</u>: For the proof see [19, 20].

In the next section we shall also use the form of the directional derivative of the cost functional (2.20) given in:

<u>Lemma 2.5</u>: <u>The directional derivative</u> $dJ(u;d)$ <u>of the cost functional</u> (2.20) <u>at</u> $u \in U$ <u>in direction</u> $d \in U$ <u>is given by</u>:

$$dJ(u;d) = \inf\{a_u'(d;y,y) - \lambda^*(u) b_u'(d;y,y) \mid y \in K(u)\} - \varepsilon \int_\Omega (u + \sum_{i=1}^{2} (\partial u/\partial x_i)) d\, dx \tag{2.21}$$

where

$$dJ(u;d) \stackrel{def}{=} \lim_{t \to 0^+} [J(u+td) - J(u)]/t$$

$$K(u) = \{\bar{y} \in V : a_u(\bar{y},\bar{y}) = \inf\{a_u(y,y) \mid y \in V,\ b_u(y,y) = 1\}\}$$

$$a'_u(d;y,y) = \int_\Omega 3u^2 d[y_{11}^2+y_{22}^2+2\nu y_{11}y_{22}+2(1-\nu)y_{12}^2]dx \qquad (2.22)$$

$$b'_u(d;y,y) = \int_\Omega d\, y^2\, dx \qquad (2.23)$$

Proof: is given in [19].

3. Finite-Dimensional Approximation of Problem (P)

In order to solve numerically problem (P) we approximate it by a finite dimensional problem (P_h). To do this we employ finite element method [4, 24]. We introduce the approximations of the set U_{ad} and the state equation (2.13).

Let us introduce the family T_h of the partitions of domain Ω depending on a discretization parameter h. The family T_h is constructed by dividing the closure $\bar{\Omega}$ of the domain $\Omega \subset R^2$ into disjoint, rectangular elements K_i, i=1,...,I, such that:

$$\bar{\Omega} = \sum_{i=1}^{I} K_i = \bigcup_{K \in T_h} K$$

We denote by h_i and $\bar{h}_i$ the diameter and the length of the smallest side of the element K_i respectively. We denote:

$$h = \max\{h_i \mid 1 \leq i \leq I\}$$

$$\bar{h} = \min\{\bar{h}_i \mid 1 \leq i \leq I\}$$

It is assumed that the discretization parameter h tends to 0 and that for all partitions T_h the following condition is satisfied [4, 24]:

$$\bar{h}/h \geq h_o > 0 \qquad h_o=\text{const}$$

Let us introduce finite dimensional spaces U_h and V_h approximating spaces $H^1(\Omega)$ and V respectively.
U_h is the space of continuous functions $\psi_j=\psi_j(.;h)$, j=1,...,N, bilinear [4] on each element K_i, i=1,...,I, V_h is the space of continuously differentiable functions $\phi_j=\phi_j(.;h)$, j=1,...,M, bicubic [4] on each element K_i, i=1,...,I.

The thickness of the plate $u \in U_{ad}$ and its displacement $y \in V$ are approximated respectively by $u_h \in U_h$ and $y_h \in V_h$ given by:

$$u_h = \sum_{i=1}^{N} q_i \psi_i \qquad y_h = \sum_{j=1}^{M} w_j \phi_j \tag{3.1}$$

where $(q_1,\ldots,q_N)=q \in R^N$ and $(w_1,\ldots,w_M)=w \in R^M$ denote vectors of coefficients.

The admissible set U_{ad} is approximated by:

$$U_{h\,ad} = \{u_h \in U_h : u_{min} \leq u_h \leq u_{max} , \quad \sum_{K \in T_h} \int_K u_h dx = c\} \tag{3.2}$$

It is assumed that constants u_{min}, u_{max}, c, and the parameter h are such that the set (3.2) is nonempty.

The bilinear forms (2.10) and (2.11) are approximated by the bilinear forms $a_{h,u_h}(y_h,\phi_h) : V_h \times V_h \to R$ and $b_{h,u_h}(y_h,\phi_h) : (L^2(\Omega) \cap V_h) \times (L^2(\Omega) \cap V_h) \longrightarrow R$ respectively. These forms are given by:

$$a_{h,u_h}(y_h,\phi_h) = \sum_{K \in T_h} \int_K u_h^3 [y_{h,11}\phi_{h,11} + y_{h,22}\phi_{h,22} + \nu y_{h,11}\phi_{h,22} + \nu y_{h,22}\phi_{h,11} + 2(1-\nu) y_{h,12}\phi_{h,12}] dx \tag{3.3}$$

$$b_{h,u_h}(y_h,\phi_h) = \sum_{K \in T_h} \int_K u_h y_h \phi_h dx \tag{3.4}$$

The cost functionals (2.17) and (2.18) are approximated by the cost functionals $I_h(u_h)$ and $R_h(u_h)$, respectively, given by:

$$I_h(u_h) = \lambda^*(u_h) = \inf\{a_{h,u_h}(y_h,y_h) \mid y_h \in V_h , \ b_{h,u_h}(y_h,y_h) = 1\} \tag{3.5}$$

$$R_h(u_h) = -0.5\, \varepsilon \sum_{K \in T_h} \int_K (u_h^2 + \sum_{i=1}^{2} (\partial u_h/\partial x_i)^2) dx \tag{3.6}$$

It is assumed that the integrals over an element $K \in T_h$ in (3.2)-(3.6) are numerically computed using an exact quadrature formula [24]. It implies that for all $h > 0$ the following condition holds:

$$U_h \subset U , \quad V_h \subset V , \quad U_{h\,ad} \subset U_{ad} \tag{3.7}$$

i.e. we use the conforming finite element method [4].

Let us denote by $z_h : V \to V_h$ a restriction operator such that for $h \to 0$:

$$\|\phi - z_h\phi\|_V \longrightarrow 0 \quad \text{for all} \quad \phi \in V \tag{3.8}$$

We shall consider a discretized optimization problem (P_h) which approximates problem (P):

$$(P_h) \quad \left| \begin{array}{l} \text{maximize the cost functional} \\ J_h(u_h) = I_h(u_h) + R_h(u_h) \qquad (3.9) \\ \text{subject to } u_h \in U_{h\,ad} \text{ and the state equation:} \\ a_{h,u_h}(y_h,\phi_h) = \lambda(u_h) b_{h,u_h}(y_h,\phi_h) \text{ for all } \phi_h \in V_h \qquad (3.10) \end{array} \right.$$

where $\phi_h = z_h\phi$, $\phi \in V$ and z_h satisfies (3.8).

$U_{h\,ad}$, $I_h(u_h)$, $R_h(u_h)$ are given by (3.2), (3.5) and (3.6) respectively.

Remark 3.1:

Since we consider finite dimensional approximation of problem (P), taking into account (3.1), problem (P_h) may be formulated in equivalent form as an optimization problem (P_q) for coefficients $q \in R^N$. In order to formulate problem (P_q) let us introduce the matrices:

$$A_h(q) = \{a_{h,u_h}(\phi_i,\phi_j)\}_{i,j=1,\dots,M}$$

$$B_h(q) = \{b_{h,u_h}(\phi_i,\phi_j)\}_{i,j=1,\dots,M} \tag{3.11}$$

$$C_h = \{ \sum_{K \in T_h} \int_K [\psi_i\psi_j + \sum_{l=1}^{2} \frac{\partial\psi_i}{\partial x_l}\frac{\partial\psi_j}{\partial x_l}]\, dx\}_{i,j=1,N}$$

and

$$d_i(h) = \sum_{K \in T_h} \int_\Omega \psi_i dx \qquad i=1,\dots,N$$

From (3.1) and (3.11) it follows that solving (3.10) for fixed u_h consists in finding pairs $(\lambda(q), w(q)) \in R \times \{R^M \setminus \{0\}\}$ satisfying general eigenvalue problem:

$$A_h(q)w = \lambda(q)\, B_h(q)w \tag{3.12}$$

where $q \in R^N$ corresponds to u_h and $w \in R^M$ corresponds to y_h according to (3.1).

$A_h(q)$ and $B_h(q)$ are called stiffness matrix and mass matrix, respectively [24].
Using (3.1) and (3.11) and taking into account the orthonormality of basis functions [4] we can described U_{had}, $I_h(u_h)$, $R_h(u_h)$ in equivalent form by:

$$U_{h\,ad} = \{q \in R^N : g_i = q_i - u_{max} \leq 0 \quad i=1,\dots,N; \quad g_i = u_{min} - q_{i-N} \leq 0 \quad i=N+1,\dots,2N; \quad g_i = \sum_{j=1}^{N} d_j(h) q_j - c = 0 \quad i=2N+1\} \tag{3.13}$$

$$I_h(q) = \lambda_h^*(q) = \inf\{(A_h(q)w,w)_{R^M} \mid w \in R^M, \ (B_h(q)w,w)_{R^M} = 1\} \tag{3.14}$$

$$R_h(q) = -0.5\ \varepsilon\ (C_h q,q)_{R^N} \tag{3.15}$$

Taking into account (3.1) and (3.11)-(3.15) we can formulate the following optimization problem (P_q) for coefficients $q \in R^N$ corresponding to problem (P_h):

(P_q) | maximize $J_h(q) = I_h(q) + R_h(q)$ subject to $q \in U_{had}$ and the state equation (3.12).

$U_{h\,ad}$, $I_h(q)$, $R_h(q)$ are given by (3.13), (3.14), (3.15) respectively.

<u>Lemma 3.1</u>: <u>There exists an optimal solution</u> $\hat{u}_h \in U_{h\,ad}$ <u>of problem</u> (P_h).

<u>Proof</u>: The set $U_{h\,ad}$ is closed and bounded in U_h, hence it is compact in U_h. From Lemma 2.3, (2.18) and (3.7) follows continuity of the cost functional (3.9) on $U_{h\,ad}$. Hence, Weierstrass theorem implies the existence of an optimal solution $\hat{u}_h \in U_{h\,ad}$ to problem (P_h).

□

We shall need the following auxiliary lemma:

<u>Lemma 3.2</u>: <u>For any</u> $u \in U_{ad}$ <u>there exists a sequence</u> $\{u_h\}$ <u>such that for</u> $h \to 0$:

$$u_h \in U_{h\,ad}, \quad u_h \longrightarrow u \quad \text{strongly in } H^1(\Omega) \tag{3.16}$$

Proof: is given in Appendix.

The convergence of approximation is shown in:

Lemma 3.3: Let $\{\hat{u}_h\} \subset U_{h\ ad}$ be a sequence of solutions to problem (P_h). There exist weak accumulation points of the sequence $\{u_h\}$ in $H^1(\Omega)$ and each such a point is a solution of problem (P).

Proof: Let $\{u_h\}$ be a sequence satisfying (3.16) and corresponding to an optimal solution $\hat{u}$ of problem (P). First we show that the sequence $\{\hat{u}_h\}$ is bounded in $H^1(\Omega)$ norm uniformly with respect to h. By definition of $\hat{u}_h$ we have:

$$J_h(\hat{u}_h) \geqslant J_h(u_h) \qquad \text{for all} \quad h > 0 \tag{3.17}$$

From (3.5), (3.6), (3.9), (3.17) we have for all $h > 0$:

$$\lambda^*(\hat{u}_h) - 0.5\varepsilon \, \|\hat{u}_h\|^2_{H^1(\Omega)} \geqslant \lambda^*(u_h) - 0.5\varepsilon \, \|u_h\|^2_{H^1(\Omega)}$$

Hence:

$$\|\hat{u}_h\|^2_{H^1(\Omega)} \leqslant \lambda^*(\hat{u}_h) - \lambda^*(u_h) + 0.5\varepsilon \, \|u_h\|^2_{H^1(\Omega)} \tag{3.18}$$

From (2.9c), (2.10),(2.11), (2.14), (2.15),(3.7) as well as from Lemmas 2.1 and 2.2 it follows that for all $h > 0$:

$$\lambda^*(u_h) \leqslant e \tag{3.19}$$

$$\|y(u_h)\|_V \leqslant e \tag{3.20}$$

where e is a generic constant independent of h. From (3.16), (3.18), (3.19) we obtain for all $h > 0$:

$$\|\hat{u}_h\|_{H^1(\Omega)} \leqslant e \tag{3.21}$$

From (3.19)-(3.21) it follows [2] that there exists a subsequence of $\{h\}$ which we denote further by $\{h\}$ such that for $h \to 0$ we obtain:

$$\begin{aligned} &\hat{u}_h \longrightarrow \bar{u} \ \text{weakly in} \ H^1(\Omega) \\ &\lambda^*(\hat{u}_h) \longrightarrow \lambda(\bar{u}) \ \text{in} \ R \\ &y(\hat{u}_h) \longrightarrow y(\bar{u}) \ \text{weakly in} \ V \end{aligned} \tag{3.22}$$

By compact imbedding $H^1(\Omega) \subset L^2(\Omega)$ and by closedness of U_{ad} in $L^2(\Omega)$ as well as by (3.7) it follows that

$$\begin{aligned} \hat{u}_h &\longrightarrow \bar{u} \quad U_{ad} \quad \text{strongly in } L^2(\Omega) \\ y(\hat{u}_h) &\longrightarrow y(\bar{u}) \quad \text{strongly in } L^2(\Omega) \end{aligned} \tag{3.23}$$

Using the same arguments as in the proof of Lemma 2.3 (see [19, 20]) and taking into account (3.7) it is easy to show that $\lambda(\bar{u})=\lambda^*(\bar{u})$, i.e. $\lambda(\bar{u})$ is the smallest eigenvalue satisfying (2.13) with $\bar{u} \in U_{ad}$.

We show that $\bar{u} \in U_{ad}$ is an optimal solution of problem (P). Taking into account (3.16), (3.17), (3.22), (3.23) we obtain for a subsequence of $\{h\}$ which we denote further by $\{h\}$ that:

$$J(\hat{u}) = \lim_{h\to 0} J_h(u_h) \leqslant \overline{\lim_{h\to 0}} J_h(\hat{u}_h) \leqslant J(\bar{u})$$

i.e. $\bar{u}=\hat{u} \in U_{ad}$ is an optimal solution of problem (P) what completes the proof.

□

4. Numerical Optimization Method

Since the smallest eigenvalue of problem (3.12) may be multiple [9, 10] (P_q) is a nonsmooth optimization problem. In order to solve it we employ the shifted penalty function method [7] combined with the Lemarechal's method [12, 15].

In order to describe these methods let us introduce a shifted penalty cost functional:

$$Q(q,p,\gamma) = I_h(q) + R_h(q) + 0.5\gamma K(q,p) \tag{4.1}$$

where

$$K(q,p) = \sum_{i=1}^{2N} \max{}^2(0;g_i-p_i) + (g_{2N+1}-p_{2N+1})^2 \tag{4.2}$$

g_i, $i=1,\dots,2N+1$, $I_h(q)$, $R_h(q)$ are given by (3.13), (3.14), (3.15) respectively. $p \in R^{2N+1}$ is a shifting vector and $\gamma \in R$, $\gamma > 0$ is a penalty coefficient. Instead of solving problem (P_q) we can solve the following equivalent problem [7]:

$$\max\{Q(q,p,\gamma) \mid q \in R^N\} \tag{4.3}$$

subject to the state equation (3.12).

The shifted penalty function method to solve problem (4.3) is described as follows [7]:

Stop 0. Let $p=p_o$, $\gamma=\gamma_o$, $l_o \in (0,1)$, $\beta \in (0,1)$, $e \in (0,1)$ be given. Set $k=0$.

Step 1. Find a solution $\hat{q}_k \in R^N$ of the problem:

$$\max\{F(q) \mid q \in R^N\} \qquad F(q) \overset{def}{=} Q(q,p_k,\gamma_k) \tag{4.4}$$

subject to the state equation (3.12).

Step 2. If $\|g_i\| \leq l_k$, $i=1,\ldots,2N+1$

then: $p_{k+1} = \beta p_k$, $\gamma_{k+1}=\gamma_k$, $l_{k+1}=el_k$;

otherwise: $p_{k+1}=p_k$, $\gamma_{k+1}=\gamma_k/\beta$, $l_{k+1}=l_k$

Set: $k=k+1$, go to Step 1.

To solve nonsmooth problem (4.4) we employ Lemarechal's method [12]. To describe briefly this method we recall the notion of subgradient. $\partial F(q)$ denote a subgradient of a Lipschitz continuous function $F(q)$ at $q \in R^N$ which is given by [12, 15]:

$$\partial F(q) = \{z \in R^N : (z,d)_{R^N} \leq dF^o(q;d) \quad \text{for all} \quad d \in R^N\}$$

where $dF^o(q;d)$ is a derivative given as in Clarke [5] by:

$$dF^o(q;d) = \limsup_{\substack{s\to 0\\ t\to 0}} [F(q+s+td) - F(q+s)]/t \tag{4.5}$$

where $s \in R^N$, $t > 0$.

The Lemarechal's method for solving problem (4.4) can be described as follows [12]:

Let $q_o \in R^N$, $u > 0$, $\eta > 0$ be given. $z_o \in \partial F(q_o)$, $i=n=0$, $m_i=0$.

Step 1. Calculate d_n being the projection of $0 \in R^N$ onto the convex hull of $\{z_{m_i},\ldots,z_n\}$. If $\|d_n\| \leq \eta$, stop.

Step 2. Calculate $\tau_n = \arg\max\{F(q+\tau d_n) \mid \tau \geq 0\}$
Set: $q_{n+1}=q_n+\tau_n d_n$.

Step 3. Determine $z_{n+1} \in \partial F(q_{n+1})$ such that

$(z_{n+1},d_n)_{R^N}=0$. Set : $n=n+1$.

Step 4. If $(z_n, q_n - q_{m_i})_{R^N} \leq u$ then go to Step 1.

Step 5. Set $i=i+1$, $m_i=n$, go to Step 1.

Mifflin has shown [15] convergence of Lemarechal's algorithm for non-convex and nondifferentiable functions which are semismooth. We show that the cost functional (4.4) is semismooth in the sense of Mifflin. Let us recall (see [14]) that $F : R^N \to R$ is called semismooth at $q \in R^N$ if: (i) F is Lipschitz continuous at $q \in R^N$, (ii) for every $d \in R^N$, every subsequence of $\{t_k\} \subset R_+$, $\{\theta_k\} \subset R^N$, $\{g_k\} \subset R^N$ such that $\{t_k\} \to 0$, $\{\theta_k | t_k\} \to 0$ and $g_k \in \partial F(q_k + t_k d_k + \theta_k)$ the sequence

$$\{(g_k, d_k)_{R^N}\}$$

has exactly one accumulation point.

We shall use the following:

Theorem 4.1 [14]: Let: $f : R^N \times R^M \to R$ whereas $B \subset R^N$ is an open set and $W \subset R^M$ is a compact set. Define:

$$I(q) = \inf\{f(q,w) \mid w \in W\} \tag{4.6}$$

If: (i) $f(.,.)$ is continuous on $B \times W$

(ii) $f(.,w)$ is differentiable on B for each $w \in W$

(iii) gradient $\nabla f(.,.)$ of f with respect to $q \in B$ is continuous and bounded on $B \times W$ then I is semismooth on B.

Proof: is given in [14].

Lemma 4.1: The cost functional (4.4) is semismooth on $U_{h\,ad}$.

Since $R_h(q)$ and $K(q,p)$ in (4.4) are smooth enough to calculate their gradients in order to show the semismoothness of (4.4) on $U_{h\,ad}$ it is enough to show that the term $I_h(q)$ of (4.4) is semismooth on $U_{h\,ad}$. To do this we have to show that $I_h(q)$ satisfies the assumptions of Theorem 4.1.

Let f, B, W be such as introduced in Theorem 4.1. Taking into account (3.1) and (3.11), as well as (3.14), Lemma 2.2 we can denote

$$f(q,w) = (A_h(q)w,w)_{R^M} / (B_h(q)w,w)_{R^M} \tag{4.7}$$

$$B = U_{h\,ad}$$

$$W = \{w \in R^M : (B_h(q)w,w)_{R^M} = 1, \quad ||w|| \leq e\}$$

where e is a given constant.
By (3.3), (3.11) and Remark 3.1 the functional (4.7) is equivalent to the form (3.3). From (2.9), (2.10), (3.7), (3.14) as well as Lemma 2.1 and 2.3 it follows continuity of (4.7) on $B \times W$. The existence and the forms (2.22) and (2.23) of gradients of the forms (3.3) and (3.4) respectively were derived in [18, 19], hence (4.7) satisfies assumption (ii) of Theorem 4.1.
By (2.9), (3.7) and Lemmas 2.1 and 2.2 it follows continuity and boundedness of gradients (2.22), (2.23) of forms (3.3), (3.4) respectively, so the gradient of (4.7) is continuous and bounded on $B \subset W$. By Theorem 4.1 we obtain the semismoothness of (4.4) on $U_{h\,ad}$.

□

Let us note that the cost functionals (3.15) and (4.2) are smooth enough to calculate their gradients. The form of the directional derivative of the functional (3.14) is given in Lemma 2.5. Hence we can calculate the directional derivative of (4.4) and use it in the Lemarechal's algorithm. By Lemma 4.1 the derivative (4.5) of the cost functional (4.4) is equal to the directional derivative of (4.4) [15].
The semismoothness of (4.4) implies [15] the convergence of the above algorithm solving problem (P_q).

Let us remark that in order to calculate the directional derivative of (4.4) and find an optimal solution of problem (P_q) we have to solve the eigenvalue problem (3.12). It is well known [17], that if two smallest eigenvalues of (3.12) are coalescing problem (3.12) becomes numerically ill-conditioned. We employ a QZ method [17] to solve this problem. This method uses unitary transformations of the matrices $A_h(q)$ and $B_h(q)$ in (3.12) which assures its numerical stability.

5. Numerical Examples

Problem (P_q) was solved numerically using the algorithms described in the preceding section. The computations were carried out for the clamped, simply supported and free plates occupying domain Ω given by:

$$\Omega = \{(x_1,x_2) \in R^2 : x_1 \in (0,a) \wedge x_2 \in (0,b)\} \tag{5.1}$$

where a,b are given real positive numbers. The boundary Γ of the domain Ω is given by:

$$\Gamma = \{(x_1,x_2) \in R^2 : (x_1=0,a \wedge x_2 \in [0,b]) \\ (x_2=0,b \wedge x_1 \in [0,a])\} \tag{5.2}$$

We do not exclude the case where the smallest eigenvalues $\lambda_h^*(q)$ satisfying (3.12) are multiple. The two smallest eigenvalues λ_1 and λ_2 satisfying (3.12) are regarded as being equal if they satisfy:

$$|\lambda_1 - \lambda_2| \leqslant \sigma \tag{5.3}$$

where σ is a given small positive number.
Lemarechal's implementation of his method was used. ODRA 1305 and NEAC Acos 900 computers were employed.
Since problem (P_q) is symmetric only one quarter or one half of the plate was used in computations. The initial thickness distribution u_o was taken to be constant. The numerical data are: $\varepsilon=10^{-5}$, $\sigma=10^{-5}$. We obtained the following results:

Table 1 contains the results of optimization for one quarter of the plate clamped along Γ_1 and simply supported along Γ_2, where Γ_1 and Γ_2 are given by:

$$\Gamma_1 = \{(x_1,x_2) \in R^2 : x_1 = 0,a \wedge x_2 \in [0,b]\}$$

$$\Gamma_2 = \Gamma \setminus \Gamma_1$$

The plate was divided into 64 elements. The numerical data are: $a=b=1.0$, $u_{min}=0.8$, $u_{max}=1.2$, $c=1.0$, $u_o=1.0$. The material concentrates in the middle of the clamped edges, near corners and in the middle of the plate. The smallest eigenvalue is double for the optimum thickness plate and is equal to 1868. The smallest eigenvalue for the initial constant plate is equal to 774.

Table 2 contains the results of optimization for one quarter of the plate simply supported along the boundary Γ given by (5.2). The plate was divided into 64 elements. The numerical data are: $a=b=1.0$, $u_{min}=0.8$, $u_{max}=1.2$, $c=1.0$, $u_o=0.8$. The material concentrates in the corners and in the middle of the plate.
The smallest eigenvalue was increased from the value 232 for the initial thickness plate to the value 432 for the optimum thickness plate.

Table 3 contains the results of optimization for the one half of the plate clamped along Γ_1 and free along Γ_3, where Γ_1 and Γ_3 are given by:

$$\Gamma_1 = \{(x_1,x_2) \in R^2 : x_1 = 0 \wedge x_2 \in [0,b]\}$$

$$\Gamma_3 = \Gamma \setminus \Gamma_1$$

The plate was divided into 56 elements. The numerical data are: a=1.0, b=3.0, u_{min}=2.0, u_{max}=4.0, c=9.0, u_o=3.0. The material concentrates near the clamped edge. The smallest eigenvalues for the initial thickness plate and the optimum thickness plate are 139 and 343 respectively.

Computing time required to obtain optimum thickness plate is about 15 min in the case of occurrence of double eigenvalues and a bit shorter in other cases. Most of this time is used for solving the eigenvalue problem (3.12). Since the mass matrix $B_h(q)$ in (3.12) is numerically near singular we computed the inverse $1|\lambda^*$ of the smallest eigenvalue. Several eigenvalues of problem (3.12) were computed in order to check whether other eigenvalues coalesce with the smallest eigenvalue at the optimal point. When the two smallest eigenvalues coalesce the problem becomes numerically ill-conditioned and computational process becomes very slowly.

Table 1. Optimum thickness for one quarter of the clamped, simply-supported plate.

x_2 \ x_1	0.000	0.125	0.250	0.375	0.500
0.500	1.18	1.10	1.11	0.99	1.09
0.375	1.20	0.93	1.17	0.94	1.03
0.250	1.04	0.80	0.86	1.16	0.95
0.125	0.80	0.80	1.20	1.08	1.11
0.000	1.09	1.20	1.16	0.98	0.86

Table 2. Optimum thickness for one quarter of the simply-supported plate.

x_2 \ x_1	0.000	0.125	0.250	0.375	0.500
0.500	0.88	0.87	0.84	0.81	0.86
0.375	0.95	1.03	0.89	0.87	0.81
0.250	1.00	1.18	0.99	0.90	0.85
0.125	1.02	1.20	1.18	1.03	0.97
0.000	0.92	1.02	1.00	0.95	0.88

Table 3. Optimum thickness for one half of the clamped, free plate

x_2 \ x_1	0.000	0.429	0.798	1.287	1.713	2.142	2.571	3.000
0.500	3.70	3.96	3.63	3.21	2.80	2.56	2.31	2.26
0.375	4.00	4.00	3.99	3.54	3.05	2.35	2.00	2.09
0.250	3.98	4.00	3.98	3.55	3.05	2.35	2.00	2.09
0.125	3.96	4.00	3.95	3.52	3.04	2.25	2.00	2.06
0.000	3.73	3.91	3.58	3.31	3.04	2.20	2.09	2.00

6. Concluding Remarks

The optimal design problem of free vibrating plates is considered in this paper. The problem consist in finding such distribution of the plate thickness which maximizes the smallest eigenvalue under the assumption that the plate volume is constant and the plate thickness is bounded. The case where the smallest eigenvalue can be multiple is covered. Finite element method is employed as an approximation method.

Convergence of the proposed approximation is shown assuming that the plate thickness is less regular than it was assumed in [13]. A non-smooth optimization method [12] is used to solve numerically the discretized problem. The convergence of this method is analyzed. Numerical results are presented.

Numerical results obtained here confirm that for optimum thickness plate the double smallest eigenvalue can occur. Since the computational process is very slow it seems necessary to accelerate it by constructing more effective numerical algorithms. An example of such algorithm can be found in [21]. This algorithm takes into account the structure of the solved problem.

7. Appendix : Proof of Lemma 3.2

We construct a sequence $\{u_h\}$ satisfying (3.16) interpolating the element $u \in U_{ad}$.

Let $\bar{z}_h : H^1(\Omega) \supset U \to U_h$ be a restriction operator such that for $h \to 0$:

$$\|\bar{z}_h u - u\|_U \longrightarrow 0 \quad \text{for all} \quad u \in U \tag{7.1}$$

and $\bar{z}_h u \in U_h$ satisfies (2.9)(c).

Let $\varepsilon = \varepsilon(h) > 0$ be such that for $h \to 0$, $\varepsilon(h) \to 0$. We define:

$$u_\varepsilon = \begin{cases} u_{max} - \varepsilon & u \geq u_{max} - \varepsilon \\ u & u_{min} + \varepsilon \leq u \leq u_{max} - \varepsilon \\ u_{min} + \varepsilon & u \leq u_{min} + \varepsilon \end{cases} \tag{7.2}$$

From (7.2) we obtain for $\varepsilon \to 0$:

$$U \ni u_\varepsilon \longrightarrow u \quad \text{strongly in} \quad H^1(\Omega) \tag{7.3}$$

By (7.1) we obtain for $h \to 0$:

$$U_h \ni \bar{z}_h u_\varepsilon \longrightarrow u_\varepsilon \quad \text{strongly in} \quad H^1(\Omega) \tag{7.4}$$

Let us define:

$$u_h = \bar{z}_h u_\varepsilon + d_h \tag{7.5}$$

where $d_h = \left[c - \int_\Omega \bar{z}_h u_\varepsilon dx\right] / \text{meas } \Omega$ and c is the constant the same as in (2.19).

From (7.2)-(7.5) as well as from the construction of operator $\bar{z}_h$ it follows that:

$$|d_h| \leqslant \varepsilon \tag{7.6}$$

$$u_{min} + \varepsilon \leqslant \bar{z}_h u_\varepsilon \leqslant u_{max} - \varepsilon \tag{7.7}$$

By (7.5)-(7.7) we obtain that $u_h \in U_{h\,ad}$. From (7.3), (7.4), (7.6) we obtain for $h \to 0$ that $\varepsilon(h) \to 0$ and $u_h \to u$ strongly in $H^1(\Omega)$.

□

<u>References</u>

[1] J.P. Aubin: Applied Functional Analysis, Wiley Interscience, New York, 1979.

[2] J.P. Aubin: Approximation of Elliptic Boundary Value Problems, Wiley Interscience, New York, New York, 1972.

[3] N.V. Banichuk: Optimization of Forms of Elastic Bodies, Nauka, Moscow, USSR, 1982 (in Russian).

[4] Ph. Ciarlet: The Finite Element Method for Elliptic Problems, North Holland, Amsterdam, Holland, 1978.

[5] F.H. Clarke: Optimization and Nonsmooth Analysis, Wiley Interscience, New York, New York, 1983.

[6] C.L. Dym, I.H. Shames: Solid Mechnaics: A Variational Approach, Mc Graw Hill, New York, New York, 1973.

[7] W. Findeisen, J. Szymanowski, A. Wierzbicki: Theory and Methods of Optimization, Polish Scientific Publisher, Warsaw, Poland, 1977 (in Polish).

[8] D.J. Gorman: Free Vibration Analysis of Rectangular Plates, North Holland, Amsterdam, Holland, 1982.

[9] E.J. Haug, J. Cea, eds.: Optimization of Distributed Parameter Structures, Sijthoff and Noordhoof, Alphen aan den Rijn, Amsterdam, Holland, 1981.

[10] E.J. Haug, B. Rousselet: Design Sensitivity Analysis in Structural Mechanics II: Eigenvalue Variations, Journal of Structural Mechanics <u>8</u> (1980), pp. 161-186.

[11] T. Kato: Perturbation Theory for Linear Operators, Springer, Berlin, Germany, 1966.

[12] C. Lemarechal: An Extension of Davidon Methods to Nondifferentiable Programming, Mathematical Programming Study 3 (1975), pp. 95-109.

[13] V.G. Litvinov: The Problem of Optimal Control of the Fundamental Frequency of a Plate of Variable Thickness, Zhurnal Vychislitelnoi Matematiki i Matematicheskoi Fiziki 19 1979 , No. 4, (in Russian).

[14] R. Mifflin: Semismooth and Semiconvex Functions in Constrained Optimization, SIAM Journal on Control and Optimization 15 (1977), pp. 959-972.

[15] R. Mifflin: A Modification and an Extension of Lemarechal's Algorithm for Nonsmooth Minimization, Mathematical Programming Study 17 (1982), pp. 77-90.

[16] S.G. Mihklin: Variational Methods in Mathematical Physics, Mir, Moscow, USSR, 1970, (in Russian).

[17] C.B. Moller, G.W. Stewart: An Algorithm for Generalized Matrix Eigenvalues Problems, SIAM Journal on Numerical Analysis 10 , (1973), pp. 241-256.

[18] A. Myśliński: Optimal Design of an Elastic Plate as a Parametric Optimization Problem, Systems Research Institute, Warsaw, Poland, Technical Report, No. 18-3/80, 1980, (in Polish).

[19] A. Myśliński: Bimodal Optimal Design of Vibrating Plates Using Theory and Methods of Nondifferentiable Optimization, Journal of Optimization Theory and Applications 46 (to appear).

[20] A. Myśliński, J. Sokołowski: Nondifferentiable Optimization Problems For Elliptic Systems, SIAM Journal on Control and Optimization 23 (to appear).

[21] A. Myśliński, Y. Sakawa: An Algorithm for Linearly Constrained Minimax Problems, in Constructive Aspects of Optimization, K. Malanowski and K. Mizukami eds., Polish Scientific Publisher, Warsaw, Poland, 1985, pp. 150-170.

[22] N. Olhoff: Optimal Design of Vibrating Rectangular Plates, Inter-

national Journal of Solids and Structures 10 1974 , pp. 93-109.

[23] N. Olhoff, J.E. Taylor: On Structural Optimization, Journal of Applied Mechanics 50 (1983), pp. 1139-1151.

[24] G. Strang, G. Fix: An Analysis of the Finite Element Method, Prentice Hall, New Yersey, 1973.

[25] V.A. Troitskii, A.A. Chwatcew: Optimization of Fundamental Eigenvalues of Elastic Thin Plates, Izwestija Akademii Nauk USSR, Mekhanika Tverdogo Tela 16 (1981), No. 4, (in Russian).

[26] J.P. Zolesio: Semiderivatives of repeated eigenvalues, in Optimization of Distributed Parameter Structures, E.J. Haug and J. Cea eds., Sijthoff and Noordhoof, Alphen aan den Rijn, Amsterdam, Holland, 1981, pp. 1457-1473.

Chapter 5

THE DESIGN OF A TWO-DIMENSIONAL DOMAIN

Antoni Żochowski

1. Introduction

There exist two approaches to the shape optimization. One of them consists in simplifying the model in every particular case, so that the parameters describing the geometry of the design appear as coefficients in a differential operator (beam or plate equations). Then the Lagrange functional is formed and the neccessary optimality conditions are obtained. Usually they consist of coupled, nonlinear, integro-partial differential boundary value problems with unknown internal boundaries. The above method may be considered as based on the neccessary conditions approach.

The alternative method is founded on the remodelling approach and, generally speaking, uses the directions of improvement in the design space. Such an approach becomes possible thanks to the recently published results [7,9,13 and many others] concerning domain differentiation technique for elliptic boundary value problems. It seemed reasonable to apply these methods to the complete system of elasticity equations, avoiding in this way the simplification in the problem formulation and nonlinearities resulting from the first approach.

In this chapter a rather simple model example of a beam is studied, but in the way that ilustrates the application of the proposed method to more general cases. In Section 2 the optimization problem is formulated and a family of admissible shapes defined. In Section 3 the existence of solution to the oryginal problem is studied together with the convergence of discretized solutions. In Section 4 the technique of domain differentiation is described and the examples of optimal shapes are given. The inadequacy of simply connected or "full" shapes is also discussed. Section 4 contains the derivation of the substitute material constants for perforated domains based on a new variational interpretation of the homogenization process. In Section 5 the formulae for these constants are used in the hierarchical method of perforation design for the beam of cellular structure and a numerical example is given. Finally at the end of Section 5 the method is additionally ilustrated by examples described by Laplace equation, the design of a heat diffusor.

The chapter contains results published already by the author elsewhere, so the proofs are usually omitted. The notations $H^n, W^{n,p}$ for Sobolev spaces are used throughout the chapter, for reference see e.g. [3]. Figures are collected at the end of the chapter.

2. The model problem

In most of the subsequent considerations we shall fix our attention on the example of the one - or two-sidedly clamped beam with the thickness uniform in the direction of z - axis, see Fig.1. Such a structure may be treated in the framework of plane elasticity theory. We shall assume the absence of gravity or internal forces in order to simplify calculations, but it does not cause the loss of generality of the discussed methods.

The part S_1 of the boundary is loaded with some force $\underset{\sim}{g}$, while S_t constitutes the design variable. The fact that S_t is free of forces has further reaching consequences because otherwise not only the computations would be more complicated, but we also would have to increase requirements concerning regularity of S_t.

The design objective is to minimize the weight of the beam ensuring simultanously that it does not bend too much, does not break and its shape is smooth enough. Now we shall give the precise formulation of the problem.

The state equation consists of the plane stress elasticity system of equations, which using matrix notation [11], takes, in terms of displacement $\underset{\sim}{u}$, the following form:

$$\begin{aligned} A^T D A \underset{\sim}{u} &= \underset{\sim}{0} \quad \text{in} \quad \underset{\sim}{\Omega}_t, \\ \underset{\sim}{u} &= \underset{\sim}{0} \quad \text{on} \quad S_o, \\ B^T D A \underset{\sim}{u} &= \underset{\sim}{g} \quad \text{on} \quad S_1, \\ B^T D A \underset{\sim}{u} &= \underset{\sim}{0} \quad \text{on} \quad S_2 \cup S_t \end{aligned} \tag{1}$$

The symbols used have the meaning:

$$A = \begin{bmatrix} \partial_x , & 0 , & \partial_y \\ 0 , & \partial_y , & \partial_x \end{bmatrix}^T \qquad\qquad B = \begin{bmatrix} n_1 , & 0 , & n_2 \\ 0, & n_2 , & n_1 \end{bmatrix}$$

$\underset{\sim}{n} = (n_1, n_2)^T$ - is the outward versor normal to $\partial\Omega_t$, $\underset{\sim}{g} = (g_1, g_2)^T$ - is the vector of load forces, D - is the 3x3-matrix of material constants.

The condition concerning the bending of the beam may be expressed as

$$||\underset{\sim}{u}|| \leqslant u_{max} \quad \text{in} \quad \Omega_t ,$$

but in order to avoid the nondifferentiability of the constraint it is approximated by

$$J_u(\underset{\sim}{u}) = (\int_{\Omega_t} \|\underset{\sim}{u}\|^p d\Omega)^{1/p} \leq u_{max}, \tag{2}$$

$p >> 1.$

The next condition is related to the yield function Y, generally depending on stresses $\underset{\sim}{\sigma}$,

$$Y(\underset{\sim}{\sigma}) \leq Y_{max}, \quad \underset{\sim}{\sigma} = DA\underset{\sim}{u}$$

where Y is a quadratic and positive function of $\underset{\sim}{\sigma}$.
For computations we have used the very simple form of Y, namely $Y(\underset{\sim}{\sigma}) = \underset{\sim}{\sigma}^T\underset{\sim}{\sigma}$ and the approximation

$$J_\sigma(\underset{\sim}{\sigma}) = (\int_{\Omega_t} Y(\underset{\sim}{\sigma})^p d\Omega)^{1/p} \leq Y_{max}, \quad p >> 1. \tag{3}$$

The constraint (3) is meant to ensure the integrity of the beam.

The last requirements concern the geometry of the structure. In the configuration from Fig.1 the shape of the domain Ω_t is defined by the function $f_t(x)$. Therefore we shall describe the set of admissible domains in terms of functions defined on the interval $[0,L]$:

$$\pi = \{\Omega_t \mid \Omega_t = \bigcup_{x \in [0,L]} [0, f_t(x)]\}$$

where

$$\text{i)} \quad 0 < y_{min} \leq f_t(x) \leq y_{max},$$

$$\text{ii)} \quad f_t \in C^1[0,L] \quad \text{and} \quad |f_t'(x)| \leq C, \tag{4}$$

iii) $\{f_t'(\cdot)\}$ forms a family of functions equicontinuous on $[0,L]$.

To make the condition iii) operational, let us recall the definition equicontinuity:

$$\forall(\varepsilon > 0) \ \exists \alpha(\varepsilon) \{|x-x'| < \alpha(\varepsilon) \Rightarrow \forall t \ |f_t(x) - f_t(x')| < \varepsilon\}$$

The above implies that the choice of the function $\alpha(\cdot)$ defines the family. Thus the set Π depends on four parameters, $\Pi = \Pi$ $(y_{min}, y_{max}, C, \alpha(\cdot))$.

Sometimes we shall use the weak formulation of (1), which may be expressed as follows:

$$a_t(\underset{\sim}{u},\underset{\sim}{v}) \triangleq \int_{\Omega_t} (A\underset{\sim}{v})^T DA\underset{\sim}{u} d\Omega = \int_{S_1} \underset{\sim}{v}^T \underset{\sim}{g} ds, \quad \forall \underset{\sim}{v} \in V_t \tag{1'}$$

where

$$V_t = \{\underset{\sim}{v}=(v_1,v_2)^T \mid \underset{\sim}{v}=\underset{\sim}{0} \text{ on } S_o \text{ and } v_1,v_2 \in H^1(\Omega_t)\}.$$

The final formulation of the minimum-weight design problem takes on the form:

$$J(\Omega_t) \overset{\Delta}{=} \int_{\Omega_t} d\Omega \to \min$$

subject to constraints

$$\Omega_t \in \Pi , \tag{5}$$

$$a_t(\underset{\sim}{u},\underset{\sim}{v}) = \int_{S_1} \underset{\sim}{v}^T \underset{\sim}{g} ds , \quad \forall \underset{\sim}{v} \in V_t ,$$

$$J_u(\underset{\sim}{u}) \leqslant u_{max}$$

and/or

$$J_\sigma(\underset{\sim}{\sigma}) \leqslant Y_{max} .$$

3. Existence of a solution and convergence of discretized designs

Proof of the existence of a solution to the optimization problem is based on a standard application of the Weierstrass theorem. Namely it is shown that in a certain topology the set of admissible designs is compact, while the goal function is continuous. To this end we shall define the set of domains L_D consisting of all measurable subset Ω of R^2 satisfying inclusions

$$D_o \subsetneq \Omega \subsetneq D_1 ,$$

where $D_o = [0,L] \times [0,y_{min}]$, $D_1 = [0,L] \times [0,y_{max}]$. L_D is endowed with the norm

$$||\Omega||^2_{L_D} = \int_{D_1} \chi^2(\Omega) d\Omega ,$$

that is the L_2 norm of the characteristic functions.

Next we shall define the subset of L_D, denoted by $U(\theta,h,r)$, containing all sets satisfying uniform cone property. The set Ω has a uniform cone property for a triple (θ,h,r), $\theta \in (0,\pi/2)$, $h > 0$, $0<r<h/2$ iff for any $\underset{\sim}{z} \in \partial\Omega$ one can find a direction $\underset{\sim}{a}_z$ such that

$$\underset{\sim}{z}_1 \in B(\underset{\sim}{z},r) \cap \Omega \Rightarrow z_1 + C(\underset{\sim}{a}_z,\theta,h) \in \Omega ,$$

where $B(\underset{\sim}{z},r)$ denotes an open ball with the center at $\underset{\sim}{z}$ and the radius r, while by C we mean the cone

$$C(\underset{\sim}{a},\theta,h) = \{\underset{\sim}{z} : \underset{\sim}{z}^T\underset{\sim}{a} > \|\underset{\sim}{z}\| \cdot \|\underset{\sim}{a}\| \cos\theta,\ \|\underset{\sim}{z}\| < h\}.$$

The following proposition is proved in [2]:

<u>Proposition 1</u>

$U(\theta,h,r)$ <u>is compact in</u> L_D . □

It is obvious that $\Pi(y_{min},y_{max},C,\alpha) \subset U(\theta,h,r)$ for suitably chosen θ,h,r. Furthermore, from the conditions (4) one obtains the next property:

<u>Proposition 2</u>

$\Pi(y_{min},y_{max},C,\alpha(\cdot))$ <u>is closed in</u> L_D □

It remains to prove the continuity of J_u and J_σ. Let $\Omega_o = [0,L] \times [0,1]$ be a reference domain. Let us assume that for sufficiently regular traction field $\underset{\sim}{g}$ the solution of the boundary value problem (1) in the domain Ω_o exists and belongs to the space $[H^2(\Omega_o)]^2$. Then the following propositions hold, [19]:

<u>Proposition 3</u>

<u>There exist constants</u> γ <u>and</u> M <u>the same for all</u> $\Omega_t \in \Pi$ <u>that</u>

i) $a_t(\underset{\sim}{u},\underset{\sim}{v}) \leq M \|\underset{\sim}{u}\|_{V_t} \|\underset{\sim}{v}\|_{V_t}$

ii) $a_t(\underset{\sim}{u},\underset{\sim}{u}) \geq \gamma \|\underset{\sim}{u}\|^2_{V_t}$ □

One of the consequences of the above proposition is the inequality [5]:

$$\|\underset{\sim}{u}\|_{V_t} \leq C_1 \|\underset{\sim}{g}\|_{[L_2(S_1)]^2} \leq C_2 \qquad (6)$$

In fact, if we assume the H^2 regularity of the solution $\underset{\sim}{u}$ in Ω_t, the inequality analogous to (6) holds for H^2 norm as well,

$$\|u\|_{[H^2(\Omega_t)]^2} \leq C_3 . \qquad (7)$$

The constants C_1, C_2, C_3 appearing in (6) and (7) are the same for all $\Omega_t \in \Pi$.

We shall also need the following property of the subset $U(\theta,h,r)$, [2].

<u>Proposition 4</u>

$U(\theta,h,r)$ <u>has the uniform extension property, i.e. for any integer</u> $m>0$ <u>and</u> $\Omega \in U(\theta,h,r)$ <u>there exists a linear and continuous extension operator</u> p_Ω <u>such that</u>

$$p_\Omega : H^m(\Omega) \to H^m(R^2)$$

<u>and</u>

$$\|p_\Omega\| \leq K(\theta,k,r,m)$$

□

We mean here that $p_\Omega(u)$ is the extension of $u \in H^m(\Omega)$ iff $p_\Omega(u)|_\Omega = u$. The symbol $K(\theta,h,r,m)$ denotes the constant depending only on the listed parameters.

From inequalities (6), (7) and Proposition 4 one may obtain the result establishing finally the existence of solution to the optimization problem (5), [19]:

<u>Proposition 5</u>

<u>If the solutions of boundary value problems</u> (1) <u>satisfy regularity condition</u> $\underset{\sim}{u} \in [H^2(\Omega_t)]^2$, <u>then functionals</u> J_u <u>and</u> J_σ <u>are continuous on</u> U <u>in</u> L_D <u>topology</u>.

□

One of the main points in the reasoning presented in this section is the assumption of H^2 regularity of solutions to (1) in Ω_t. There are examples of smooth domains and traction fields for which displacements do not belong to $H^2 \times H^2$ and conversly: for nonsmooth domains there may exist even analytical solutions. That is why the regularity of $\underset{\sim}{u}$ is assumed, instead of stating sufficient conditions in terms of the data (boundary $\partial\Omega$, traction field and displacement field). Such sufficient conditions are difficult to be formulated and would be unduly restrictive.

In numerical experiments the boundary value problem was solved using linear finite elements on triangules. The conditions (4) easily enable to construct the triangulation in such a way that

$$\frac{h}{\rho} \leq K_1(y_{min}, y_{max}, C) \quad \text{for all} \quad \Omega \in \Pi \tag{8}$$

where h - is the diameter of the triangle, ρ - is the radius of the inscribed circle. From (8) one obtains also another inequality

$$\text{area } (\Omega_h \div \Omega) \leq K_2 h \tag{9}$$

where $\div$ means symmetric difference and Ω_h denotes the union of all triangles. Both (8) and (9) allow to prove [19]:

Proposition 6

If the stiffness matrix is computed without integration errors, then the finite element solution satisfies the following error estimate uniformly in Π

$$\|\underset{\sim}{u}-\underset{\sim}{u}^h\|_{V_1(\Omega_h)} \leqslant C \cdot h^{1/2} \tag{10}$$

□

The error in evaluating the functional J_u may be obtained immidiately from (10) via embedding theorem and has also order $h^{1/2}$.

With regard to J_σ, one must consider the error assessment in $W^{1,p}$ norm. In general, the accuracy will deteriorate with increasing p. For example, if we neglect the error in approximating the boundary, the right hand side of (10) would be of the order $O(h)$, while $W^{1,p}$ norm would tend to $O(h|\ln h|)$ with $p \to \infty$.

In order to formulate the discrete optimization problem, the family Π^h of admissible domains Ω_h must be defined. It is easily done [19] by expressing conditions (4) in terms of the constraints imposed on the movements of the boundary nodes. Next we choose a decreasing sequence $\{\varepsilon_n\}$, $\varepsilon_n \to 0$ and for each ε_n we find the corresponding mesh parameter h_n such that the functionals J, J_u, J_σ are computed with accuracy ε_n. Then the sequence of discrete minimization problems $P(\varepsilon_n)$ is formulated:

$$J(\Omega_{h_n}) \to \min$$

with constraints:

$$\Omega_{h_n} \in \Pi^{h_n},$$

$$\text{discrete state equation,} \tag{11}$$

$$J_u(\underset{\sim}{u}^{h_n}(\Omega_{h_n})) \leqslant u_{max} + \varepsilon_n,$$

and/or

$$J_\sigma(\underset{\sim}{u}^{h_n}(\Omega_{h_n})) \leqslant Y_{max} + \varepsilon_n.$$

Without relaxing constraints in (11) it might happen that the shape admissible for the original problem would not be admissible for the discrete one.

Of course, it is impossible to prove the convergence of optimal design to any particular shape. However, the following results holds [19]:

Proposition 7

The sequence of solutions to discrete problems $P(\varepsilon_n)$ converges to the optimal solution of the original problem in the sense that

$$J(\Omega^*_{h_n}) \to J^*$$

where star denotes optimality. □

4. The method of solution and examples of optimal shapes

Any efficient algorithm applied to the optimization problem (5) requires computation of the gradients of both the goal functional and constraints with respect to the design variable - in our case the shape of the domain.

Let $s(x)$ be a given function defined on $[0,L]$ and ϕ_τ be a transformation defined on $[0,L] \times R$

$$\phi_\tau : \begin{array}{l} x \to x \\ y \to y + \tau \cdot y \dfrac{s(x)}{f_t(x)} \end{array}$$

The disturbed domain $\Omega_{t+\tau} = \phi_\tau(\Omega_t)$ is obtained from Ω_t by moving the boundary S_t in the direction of the y axis on the distance $\tau \cdot s(x)$. Let $\underset{\sim}{u}_\tau$ be the solution to (1) in $\Omega_{t+\tau}$. The function $\tau \to \underset{\sim}{u}_\tau$ is locally differentiable in Ω_t if the restriction of $\underset{\sim}{u}_\tau$ to any open set $\Omega'_t \subsetneq \Omega_t$ is differentiable. Then the local derivative $\partial \underset{\sim}{u}_\tau / \partial\tau |_{\tau=0}$ is defined in the whole domain Ω_t by its restriction to any Ω'_t. Formally we may write in any Ω'_t

$$\underset{\sim}{u}' = \frac{\partial \underset{\sim}{u}_\tau}{\partial \tau}\bigg|_{\tau=0} = \lim_{h\to 0} \frac{1}{h}(\underset{\sim}{u}_\tau - \underset{\sim}{u}_o) \quad . \tag{12}$$

The differentiability of $\underset{\sim}{u}_\tau$ in $\Omega'_t \subsetneq \Omega_t$ has been proved in [13] as well as the fact that if $\underset{\sim}{u} \in [W^{m,p}(\Omega_t)]^2$ and $s \in C^1[0,L]$ then $\underset{\sim}{u}' \in [W^{m-1,p}(\Omega_t)]^2$. Moreover if F is a differentiable function, then the following relation holds

$$\frac{d}{d\tau} \int_{\Omega_{t+\tau}} F(\underset{\sim}{u}_\tau)\, d\Omega \bigg|_{\tau=0} = \int_{\Omega_t} D_{\underset{\sim}{u}} F(\underset{\sim}{u}) \cdot \underset{\sim}{u}'\, d\Omega + \int_{S_t} F(\underset{\sim}{u}) n_2 s\, dS \tag{13}$$

In [13] it is also shown that $\partial/\partial\tau$ and $\partial/\partial x$, $\partial/\partial y$ commute and therefore (13) may be applied to the weak form of the state equation (1') yielding

$$\int_{\Omega_t} (A\underset{\sim}{v})^T DA\underset{\sim}{u}'\, d\Omega + \int_{S_t} (Av)^T DA\underset{\sim}{u} s n_2\, dS = 0 \tag{14}$$

Using (14) and introducing the adjoint equation

$$\int_{\Omega_t} (A\underset{\sim}{v})^T DA\underset{\sim}{w} d\Omega = \int_{\Omega_t} D_{\underset{\sim}{u}} F(\underset{\sim}{u}) \cdot \underset{\sim}{v} d\Omega \qquad \forall \underset{\sim}{v} \in V_t \tag{15}$$

we obtain

$$\frac{d}{d\tau} \int_{\Omega_{t+\tau}} F(\underset{\sim}{u}_\tau) d\Omega \Big|_{\tau=0} = \int_{S_t} \left[-(A\underset{\sim}{w})^T DA\underset{\sim}{u} + F(\underset{\sim}{u})\right] sn_2 dS \ . \tag{16}$$

In our case the adjoint equations have the forms: for J_u,

$$\int_{\Omega_t} (A\underset{\sim}{v})^T DA\underset{\sim}{w} d\Omega = \int_{\Omega_t} p \, \|\underset{\sim}{u}\|_p^{p-2} \, \underset{\sim}{u}^T \underset{\sim}{v} d\Omega \ , \qquad \forall \underset{\sim}{v} \in V_t \ , \tag{17}$$

and respectively for J_σ,

$$\int_{\Omega_t} (A\underset{\sim}{v})^T DA\underset{\sim}{z} d\Omega = \int_{\Omega_t} p |Y(\underset{\sim}{\sigma})|^{p-1} \nabla Y(\underset{\sim}{\sigma}) DA \, \underset{\sim}{v} \, d\Omega \ , \qquad \forall \underset{\sim}{v} \in V_t \ , \tag{18}$$

where $\nabla Y(\xi_1,\xi_2,\xi_3) = (\partial_1 Y, \partial_2 Y, \partial_3 Y)$.

The formula (16) holds also for the finite element approximant $\underset{\sim}{u}^h$ of $\underset{\sim}{u}$. It was derived in [19] from the fact that the stiffness matrix and hence also $\underset{\sim}{u}_h$ are smooth with respect to the small displacements of triangulation nodes. Therefore the discretized versions of (14)÷(18) may be also obtained by applying the above reasoning to the discrete state equation.

For solving the discretized optimal design problem the algorithm of consecutive linearisation is used, for details see [15], and in some cases also the modified version of the Pshenichnyi method [10], [6]. The full account of numerical experiments is given in [15], [16], [17], [19]. Below only some of the more interesting optimal shapes are discussed.

Let us specify the data used in numerical experiments leading to these optimal shapes.

The constraints in all cases are related to the reference shape $f_o \equiv 1$ on $[0,L]$, L=5, which served also as a starting design in the optimization process. For such a reference shape the maximal displacement and the maximal value of the yield function were computed and substituted for u_{max} and Y_{max} respectively. The remaining data were $\lambda=\mu$, $y_{min}=0.5$, $y_{max}=2$.

In Fig.2 the minimum-weight designs of one-sidedly clamped, uniformly loaded beam are shown, in two versions: displacement constrained and yield constrained.

In Fig.3 the case of two-sidedly clamped beam is displayed, for concentrated and distributed loads and various types of constraints.

Note that the shapes obtained are optimal only with respect to the admissible set Π. If we analyse the distribution of yield function inside the material, we can see great differences in its value, see e. g. in Fig.4 the case of the yield constrained "bridge". It indicates that in some places the weaker material could be used. The strength of material may be controlled by using perforation, which would decrease the weigth of the structure. Summing up, by admitting the perforated domains in the optimization process we may expect further gains in comparison to the designs depicted above.

5. Substitute coefficients by variational interpretation of homogenition

The perforated domain may be considered as consisting of a finite number of cells, each cell containing a hole. In principle finding the optimal perforation, i.e. finding optimal shapes of holds, is a case of optimal shape design discussed in the previous sections. In practice such an approach is infeasible and therefore we propose a hierarchical procedure. At the upper level the cells are treated as full and their properties are represented by substitute material coefficients. These coefficients serve as the design variables. At the lower level the shape of holes in all cells are chosen in such a way that the substitute coefficients calculated for the perforated cell approximate the coefficients given by the upper level. Because the substitute coefficients are the functions of the hole shape, the optimal shape design problem must be solved for each cell at the lower level.

In this section we shall derive formulae for substitute coefficients using variational method [18]. Let the domain $\Omega \in R^2$ be union of a certain number of small cells ω_i, each of the size h. All cells are identical and contain a smooth hole. We shall denote the cell without a hole by $\overline{\omega}_i$, its outer boundary by γ_i, and the boundary of the hole by τ_i, see Fig.5. The variational interpretation exploits the analogy between the division of the domain Ω into cells ω_i and the partitioning used in finite element approximation. We shall consider the following problem

$$
\begin{aligned}
A^T D A \underset{\sim}{u} &= 0 \quad \text{in} \quad \Omega, \\
\underset{\sim}{u} &= \underset{\sim}{u}_b \quad \text{on} \quad \partial\overline{\Omega}, \\
B^T D A \underset{\sim}{u} &= 0 \quad \text{on} \quad \bigcup_i \tau_i .
\end{aligned}
\tag{19}
$$

The state of the system (19) is described by the 3-dimensional strain vector $\underset{\sim}{\varepsilon}=A\underset{\sim}{u}$. Let us denote

$$\underset{\sim}{\rho}_1 = (x_1,0), \quad \rho_2 = (0,x_2) \quad \rho_3 = \frac{1}{2}(x_2,x_1) \tag{20}$$

These functions satisfy equations

$$A\underset{\sim}{\rho}_1=(1,0,0), \quad A\underset{\sim}{\rho}_2=(0,1,0), \quad A\underset{\sim}{\rho}_3=(0,0,1) \tag{21}$$

and therefore might form a basis for a constant deformation finite element approximation.

Now we shall define auxiliary elasticity problems in a cell ω (all ω_i are identical):

$$\begin{aligned} A^T DA\underset{\sim}{w}_i &= 0 \quad \text{in} \quad \omega, \\ \underset{\sim}{w}_i &= \rho_i \quad \text{on} \quad \gamma, \\ B^T DA\underset{\sim}{w}_i &= 0 \quad \text{on} \quad \tau, \end{aligned} \tag{22}$$

$i=1,2,3$. To (22) corresponds the bilinear form $a_D(.,.)$,

$$a_D(\underset{\sim}{u},\underset{\sim}{v}) = \int_{\underset{\sim}{\omega}} (A\underset{\sim}{u})^T DA\underset{\sim}{v}\,d\Omega,$$

but we shall need also the form corresponding to the full cell $\bar{\omega}$:

$$\bar{a}_D(\underset{\sim}{u},\underset{\sim}{v}) = \int_{\bar{\omega}} (A\underset{\sim}{u})^T DA\underset{\sim}{v}\,d\Omega.$$

Next we form the linear combinations

$$\underset{\sim}{\rho} = \alpha_1\underset{\sim}{\rho}_1 + \alpha_2\underset{\sim}{\rho}_2 + \alpha_3\underset{\sim}{\rho}_3 \; ; \quad \underset{\sim}{w} = \alpha_1\underset{\sim}{w}_1 + \alpha_2\underset{\sim}{w}_2 + \alpha_3\underset{\sim}{w}_3 \; . \tag{23}$$

Note that $\underset{\sim}{w}$ is defined in ω only, not in $\bar{\omega}$.

<u>Proposition 8</u> [18]

<u>The following values of substitute material coefficients</u> $\hat{D}=[\hat{d}_{ij}]$,

$$\hat{d}_{ij} = \frac{1}{|\bar{\omega}|}\, a_D(\underset{\sim}{w}_i,\underset{\sim}{w}_j), \quad i,j=1,2,3 \tag{24}$$

<u>minimize the functional</u>

$$J(\tilde{D}) = [\bar{a}_{\tilde{D}}(\underset{\sim}{\rho},\underset{\sim}{\rho}) - a_D(\underset{\sim}{w},\underset{\sim}{w})]^2 \tag{25}$$

<u>for all</u> $\alpha_1,\alpha_2,\alpha_3$, <u>rendering it zero</u>. □

If the cells were full, the solution to (19) might be approximated on them by functions of the form similar to ρ, (20), (23). Because of the existence of the holes, functions of $\underset{\sim}{w}$ type are used in approximation, and they coincide with $\underset{\sim}{\rho}$ only on the boundary γ. Then, the substitute coefficients are calculated that give the same value of bilinear form for both $\underset{\sim}{\rho}$-type and $\underset{\sim}{w}$-type approximants,

$$\bar{a}_{\hat{D}}(\underset{\sim}{\rho},\underset{\sim}{\rho}) = a_D(\underset{\sim}{w},\underset{\sim}{w}). \tag{26}$$

In other words, the discrete equations obtained from w-type approximation in the perforated domain, and from $\underset{\sim}{\rho}$-type approximation in the full domain but with different matrix of material coefficients (namely $\hat{D}$), are identical if one uses formulae (24).

The variational interpretation given by Proposition 8 suggests some other interesting possibilities. Often we are interested in the external properties of perforated cells, that is the relation between displacement $\underset{\sim}{u}$ and traction (or load force) $\underset{\sim}{t}=B^TDA\underset{\sim}{u}$ on the boundary γ only.

For example we may try to match the tractions on γ, when the displacements on the same part of the boundary are given.

Such an approach leads to the minimization of the different functional.

<u>Proposition 9</u> [18]

<u>Material constants</u> $\bar{D}=[\bar{d}_{ij}]$ <u>given by</u>

$$\bar{d}_{ij} = \int_\gamma n_i t_i^j d\gamma \,/\! \int_\gamma n_i^2 d\gamma\,,\quad i=1,2\,,\quad j=1,2,3 \tag{27}$$

$$\bar{d}_{3j} = \int_\gamma (n_2 t_1^j + n_1 t_2^j)\, d\gamma \,/\! \int_\gamma (n_1^2+n_2^2)\, d\gamma\,,\quad j=1,2,3$$

<u>minimize the functional</u>

$$J(\tilde{D}) = \int_\gamma \|B^T\tilde{D}A\underset{\sim}{\rho} - B^TDA\underset{\sim}{w}\|^2 d\gamma \tag{28}$$

<u>for all</u> $\alpha_1,\alpha_2,\alpha_3$. □

Here $\underset{\sim}{t}^j=(t_1^j,t_2^j)^T$ is the traction corresponding to $\underset{\sim}{w}_j$, $\underset{\sim}{t}^j=B^TDA\underset{\sim}{w}_j$, $j=1,2,3$.

Let us stress here that the variational interpretation is not only the way of expressing the homogenization procedure. It was instrumental in obtaining useful extension formulated as Proposition 9.

Now we shall ilustrate the variational approach with some simpler problem, namely the Laplace-type boundary value problem,

$$\nabla(A \cdot \nabla u) = 0 \quad \text{in} \quad \Omega, \qquad (29)$$
$$u = g \quad \text{on} \quad \partial\bar{\Omega},$$
$$\frac{\partial u}{\partial n_A} = 0 \quad \text{on} \quad \bigcup_i \tau_i .$$

A denotes here 2 x 2 symmetric positive matrix. Analogically to (20) the functions ρ_1, ρ_2

$$\rho_1 = x_1 , \quad \rho_2 = x_2 , \quad \nabla\rho_1 = (1,0), \quad \nabla\rho_2 = (0,1)$$

are defined and auxiliary boundary value problems are constructed:

$$\nabla(A \cdot \nabla w_i) = 0 \quad \text{in} \quad \omega,$$
$$w_i = \rho_i \quad \text{on} \quad \gamma, \qquad (30)$$
$$\frac{\partial w_i}{\partial n_A} = 0 \quad \text{on} \quad \tau, \ i=1,2.$$

Now let us denote

$$\rho = \alpha_1\rho_1 + \alpha_2\rho_2 , \quad w = \alpha_1 w_1 + \alpha_2 w_2 .$$

The following two propositions hold, [18].

Proposition 10

The matrix $\hat{A} = [\hat{a}_{ij}]$ given by

$$\hat{a}_{ij} = \frac{1}{|\bar{\omega}|} \int_\omega (\nabla w_i)^T \cdot A \cdot \nabla w_j \, d\Omega, \quad i,j=1,2 \qquad (31)$$

minimizes the functional

$$J(\tilde{A}) = \left| \int_{\bar{\omega}} (\nabla\rho)^T \cdot A \cdot \nabla\rho \, d\Omega - \int_\omega (\nabla w)^T \cdot A \cdot \nabla w \, d\Omega \right|^2 \qquad (32)$$

for all α_1, α_2, rendering it 0. □

Proposition 11

The matrix $\bar{A} = [\bar{a}_{ij}]$ given by

$$\bar{a}_{ij} = \int_\gamma n_i \frac{\partial w_j}{\partial n_A} d\gamma \Big/ \int_\gamma n_i^2 d\gamma , \quad i,j=1,2 \qquad (33)$$

minimizes the functional

$$J(\tilde{A}) = \int_\gamma \left(\frac{\partial\rho}{\partial n_{\tilde{A}}} - \frac{\partial w}{\partial n_A} \right)^2 d\gamma \qquad (34)$$

for all α_1, α_2. □

The formulae (24) and (31) are well known, see e.g. [12], [20], [4], and have been obtained there via asymptotic expansion of the solution u with respect to the cell size h. However, the similarity to the finite element method (FEM) scheme shown here gives intuitive understanding of the homogenization process. The formulae (27), (33) are new.

In order to check the applicability of substitute coefficients in the hierarchical optimization outlined at the begining of this section, we have tested the quality of approximation which they allow to obtain. In the first case the Laplace equation in two types of cells has been solved. The first cell contained the circular hole, r=0.25h, and the second the elliptical hole, a=0.25h, b=0.125h, $\alpha=\Pi/4$, see Fig.6. The original material was isotropic, A=I. The following values of homogenized coefficients were obtained:

I : $\hat{a}_{11}=\hat{a}_{22}=0.690,\ \hat{a}_{12}=0;\ \bar{a}_{11}=\bar{a}_{22}=0.720,\ \bar{a}_{12}=0$

II : $\hat{a}_{11}=\hat{a}_{22}=0.828,\ \hat{a}_{12}=0.061;\ \bar{a}_{11}=\bar{a}_{22}=0.876,\ \bar{a}_{12}=0.058$

The test problem was as follows:

$$\nabla(I\cdot\nabla u)=0 \quad \text{in} \quad \omega,\ u=0 \quad \text{on} \quad \gamma_o,\ \frac{\partial u}{\partial n_I}=0 \quad \text{on} \quad \tau,$$

$$\frac{\partial u}{\partial n_I}=\begin{cases}-1 & \text{on } \gamma_1 \text{ (constant flux heating)},\\ u & \text{on } \gamma_2 \text{ (cooling)}.\end{cases}$$

It has been solved exactly using boundary elements method (BEM) and then in the full cell $\bar{\omega}$ for three sets of coefficients: oryginal A=I, $\hat{A}$ and $\bar{A}$. The L_2 errors of values of u along $\gamma_1\cup\gamma_2$ were accordingly:

I : $e_I^2=0.051,\quad \hat{e}^2=0.017,\quad \bar{e}^2=0.011;$

II : $e_I^2=0.028,\quad \hat{e}^2=0.006,\quad \bar{e}^2=0.003.$

In Fig.7 the relative error for case I is shown. In addition elasticity example was solved for case II with the load force replacing heat flux on γ_1 and unloaded γ_2 part of the boundary. The original material had parameters E=1 (Young modulus in relative units), $\nu=0.2$ (Poisson constant). Because the BEM program for anisotropic medium was not available, the homogenized equation has been solved with FEM (bilinear elements). Using two different methods might distort the comparability of exact and homogenized solutions, masking with systematic error differences between the methods.

The coefficients resulting from homogenization were:

original: $d_{11}=d_{22}=1.042$, $d_{33}=0.417$, $d_{12}=0.208$, $d_{13}=d_{23}=0$,

homogenized: $\hat{d}_{11}=\hat{d}_{22}=0.922$, $\hat{d}_{33}=0.358$, $\hat{d}_{12}=0.198$, $\hat{d}_{13}=\hat{d}_{23}=0.052$.

$\bar{d}_{11}=\bar{d}_{22}=0.934$, $\bar{d}_{33}=0.432$, $\bar{d}_{12}=0.265$. $\bar{d}_{13}=\bar{d}_{23}=0.038$.

The error: $e^2_{orig}=0.200$, $\hat{e}^2=0.085$, $\bar{e}^2=0.093$.

In Fig.8 the corresponding solutions are compared.

The main objective of these numerical experiments was to test new formulae (27), (33) against (24) and (31). It turns out that in the special case when we are interested only in the extermal bahaviour of the cells (observation on the boundary) the new expressions for substitute coefficient give better or at least not worse results. Both sets of formulae allow to approximate the response of the perforated cell rather well.

6. Hierarchical method of optimization perforation design

In principle, the optimal design of a beam should consist in simultaneous changing both the domain $\bar{\Omega}$ and the shapes of holes. We shall restrict our consideration to the simpler case, where only perforation constitutes the design parameter, $\bar{\Omega}$ being constant. The model problem will be as follows (see Fig.9):

$$J(P) = \int_{\Omega(P)} d\Omega \to \min$$

$$A^T D A \underset{\sim}{u} = 0 \quad \text{in} \quad \Omega(P), \tag{35}$$

$$\underset{\sim}{u} = 0 \quad \text{on} \quad \Gamma_o, \quad B^T D A \underset{\sim}{u} = \begin{cases} (0,-1)^T & \text{on } \Gamma_1, \\ 0 & \text{on } \Gamma_2 \cup \Gamma_3 \end{cases}$$

$$J_u(\underset{\sim}{u}) = \int_{\Gamma_1} \|\underset{\sim}{u}\|^p \, ds \leqslant u^p_{max}$$

The domain $\Omega(P)$ is divided into identical cells, containing holes. The parameters defining shapes of the holes are denoted by P and called perforation parameters. In the example solved numericaly the holes are circular, so $P=(r_1,\ldots,r_N)$, where r_i - is the diameter of the i-th hole. The problem (35) describes the minimum weight design with constraint on the displacement along the upper edge.

The main difficulty lies in computing derivatives of J_u with respect to P. It will be done in two steps. Let us assume that each cell has been replaced by a full one, with substitute material coefficients. Furthermore, let $\delta\hat{D}$ denote the variation of the matrix $\hat{D}$ of substitute coefficients ($\hat{D}$ can assume different values at each cell). We may repeat the reasoning leading to (14)÷(16) and obtain

$$\delta J_u(\underset{\sim}{u}) = \int_\Omega (A\underset{\sim}{w})^T \cdot \delta\hat{D} \cdot A\underset{\sim}{u}d\Omega \tag{36}$$

where $\underset{\sim}{w}$ is a solution of the adjoint equation:

$$\int_\Omega (A\underset{\sim}{w})^T \cdot \hat{D} \cdot A\underset{\sim}{w}\, d\Omega = \int_{\Gamma_1} p\, ||\underset{\sim}{u}||^{p-2}\underset{\sim}{u} \cdot \underset{\sim}{\rho}dS \tag{37}$$

for all $\underset{\sim}{\rho} \in V(\bar{\Omega}) \overset{\Delta}{=} \{(\rho_1,\rho_2)\,|\,\rho_1,\rho_2 \in H^1(\bar{\Omega}),\ \rho_1,\rho_2=0 \text{ on } \Gamma_o\}$.

In the second step we shall consider the relation between $\delta\hat{A}$ and the perforation parameters. Let us concentrate on a given cell ω (we drop subscript i for clarity). We shall use the technique of differentiation with respect to the shape of the hole τ, described in section 4, which applied to (24) yields (definitions of w_i - see (22))

$$\hat{d}'_{ij} = \frac{1}{|\bar{\omega}|}\Big[\int_\omega ((A\underset{\sim}{w}_i)^T DA\underset{\sim}{w}'_j + (A\underset{\sim}{w}'_i)^T DA\underset{\sim}{w}_j)d\Omega + \int_\tau (A\underset{\sim}{w}_i)^T DA\underset{\sim}{w}_j(\underset{\sim}{s}\cdot\underset{\sim}{n})dS\Big]. \tag{38}$$

Then the equations for adjoint variables $\underset{\sim}{p}_j$ are defined, j=1,2,3.

$$\int_\omega (A\underset{\sim}{p}_j)^T DA\underset{\sim}{\rho}d\Omega = \int_\omega (A\underset{\sim}{w}_j)^T DA\underset{\sim}{\rho}d\Omega \tag{39}$$

for all $\underset{\sim}{\rho} \in \Psi(\omega) = \{\rho \in [H^1(\omega)]^2,\ \underset{\sim}{\rho}=0 \text{ on } \gamma\}$.

Differentiating the weak forms of equations (22) with respect to the shapes of holes and making appropriate cross-substitutions one obtains

$$\hat{d}'_{ij} = \frac{1}{|\bar{\omega}|}\int_\tau [(A\underset{\sim}{w}_i)^T DA\underset{\sim}{w}_j - (A\underset{\sim}{w}_i)^T DA\underset{\sim}{p}_j - (A\underset{\sim}{w}_j)^T DA\underset{\sim}{p}_i](\underset{\sim}{s}\cdot\underset{\sim}{n})dS \tag{40}$$

The vector $\underset{\sim}{s}$ denotes in (38) and (40) the direction of movement of the points on the boundary τ. If one restrict the admissible shapes of holes, for example to ellipses with diameters a,b and angle of incluation α, see Fig.6, or circles of diameters r, then it is posible to compute $\underset{\sim}{s}\cdot\underset{\sim}{n}$ analyticaly and to obtain $\partial\hat{d}_{ij}/\partial a$, $\partial\hat{d}_{ij}/\partial b$, $\partial\hat{d}_{ij}/\partial\alpha$ or $\partial\hat{d}_{ij}/\partial r$ from (40).

Combining (40) with (36) gives the relation between δJ_u and the variations of the hole parameters in the cells.

The relations outlined above were tested by solving the numerical example for problem (35). The beam consisted of 21 square cells in 3x7 configuration, see Fig.9. Each cell contained a circular hole, with diameters satisfying, $0.1h \leq r_i \leq 0.3h$. The reference beam had all holes identical, $r_i = 0.2h$, $i=1,\dots,21$. For such a beam the maximal displacement was calculated and substituted for u_{max} in (35).

The weight of the optimal design was also compared to this reference beam. Again the linearisation optimization algorithm has been used. The gain in the weight of the removed material (area of all holes) was about 88%, while the decrease of the weight of the beam was 13%. The results, Fig.9. agree with the distribution of the yield function, Fig.4. In places where stresses are bigger the holes are smaller.

The further ilustration the method is given by an example of the heat (Laplace) equation. The geometry of the domain is shown in Fig.10. Part Γ_o of the boundary is kept at constant temperature, part Γ_1 is heated with constant flux, parts Γ_2 and Γ_3 are cooled. The goal consists in minimizing the L_2 norm of the temperature over Γ_3

$$J(P) = \frac{1}{2} \int_{\Gamma_3} u^2 dS \tag{41}$$

subject to the state equation

$$\nabla(A \cdot \nabla u) = 0 \quad \text{in} \quad \Omega(P),$$

$$u = 0 \quad \text{on} \quad \Gamma_o,$$

$$\frac{\partial u}{\partial n_A} = \begin{cases} -1 & \text{on} \quad \Gamma_1 \\ u & \text{on} \quad \Gamma_2 \cup \Gamma_3 \end{cases}$$

Let $\delta\hat{A}$ be a variation of A and let $V = \{\rho \in H^1(\bar{\Omega}), \rho = 0 \text{ on } \Gamma_o\}$. Similarly as in (36), (37) we introduce the adjoint equation

$$\int_{\Omega} (\nabla\rho)^T \hat{A} \cdot \nabla w d\Omega + \int_{\Gamma_2 \cup \Gamma_3} w\rho dS = -\int_{\Gamma_3} u\rho dS, \quad \forall \rho \in V \tag{42}$$

and calculate the corresponding variation of J:

$$\delta J = \int_{\Omega} (\nabla w)^T \cdot \delta A \cdot \nabla u d\Omega \tag{43}$$

The relation between $\delta\hat{A}$ and parameters of the shapes of holes may be obtained in the same way as in (39), (40). Let $\Psi = \{\rho \in H^1(\omega), \rho = 0 \text{ on } \gamma\}$ and let us introduce the adjoint equations for p_j, $j=1,2$ (w_i - see (30)):

$$\int_{\omega} \nabla p_j A \cdot \nabla\rho d\Omega = \int_{\omega} \nabla w_j A \cdot \nabla\rho d\Omega, \quad \forall \rho \in \Psi . \tag{44}$$

Then, the following holds

$$a'_{ij} = \frac{1}{|\bar{\omega}|} \int_{\tau} \left[\nabla w_i A \nabla w_j - \nabla w_i A \nabla p_j - \nabla w_j A \nabla p_i \right] (\underset{\sim}{s} \cdot \underset{\sim}{n}) dS \tag{45}$$

It may be noted, that while the original equations for w_j are homogeneous, the adjoint ones (44) are not and therefore cannot be easily solved using BEM. It creates some disadvantage, because changing shapes of the holes requires redesigning of the triangulation.

In the case of the second set of the substitute coefficients $\bar{A}=[\bar{a}_{ij}]$, (33), differentiating the equations defining w_j with respect to the domain we obtain

$$\int_{\omega} \nabla w'_j A \nabla \rho d\Omega + \int_{\tau} \nabla w_j A \nabla \rho (\underset{\sim}{s} \cdot \underset{\sim}{n}) dS = 0 , \quad \forall \rho \in \Psi \tag{46}$$

Simultanuously we define the adjoint problems: find $q_i \in \Psi$

$$\int_{\omega} \nabla \rho \, A \, \nabla q_i d\Omega = \int_{\gamma} n_i \frac{\partial \rho}{\partial n_A} dS / \int_{\gamma} n_i^2 dS \tag{47}$$

for all $\rho \in \Psi$, i=1,2.

Combining (46) and (47) gives

$$\bar{a}'_{ij} = \int_{\gamma} n_i \frac{\partial w'_j}{\partial n_A} dS / \int_{\gamma} n_i^2 dS = - \int_{\tau} \nabla w_j A \nabla q_i (\underset{\sim}{s} \cdot \underset{\sim}{n}) dS \tag{48}$$

The equation (47) may be easily solved using BEM. The same reasoning applies in the case of the substitute elastic coefficients $\bar{d}_{ij}$, (27). Note that we have chosen the case of Laplace equation for ilustration only in order to avoid complicated expressions.

In general it seems that coefficients $\bar{d}_{ij}$ and $\bar{a}_{ij}$ are slightly easier to implement in the optimization procedure. Their advantage might also lie in higher accuracy for problems with the goal functions defined on the boundary of the domain.

Two numerical examples has been solved for Laplace equation. In each case the body was built of 25 square cells. In the first case the holes were assumed to be circular, $0{,}1h \leqslant r_i \leqslant 0.3h$. The optimal perforation is shown in Fig.10. The decrease of goal functional turned out to be quite substantial, $J_{opt}/J_{init}=0.57$ in comparison with the design with maximal holes, $r_i=0.34h$, i=1,...,25. In agreement with physical intuition a cone of smaller conductivity appears, directing the heat flow up-and down-wards. In the second example, the holes were elliptic, with constant a=0.25h, b=0.125h, and variable α, see Fig.6. The problem was also a little different,

$$J(P) = \frac{1}{2} \int_{\Gamma_1} u^2 dS$$

$$\frac{\partial u}{\partial n_A} = \begin{cases} u & \text{on} \quad \Gamma_0 \cup \Gamma_2 \cup \Gamma_3, \\ -1 & \text{on} \quad \Gamma_1, \end{cases}$$

It can be interpreted as the optimal design of a radiator. This time the gains were smaller, $5 \div 10\%$, depending on the initial configuration. Such a problem has two equivalent solutions, symmetrical with respect to the central horizontal line in Fig.11. Again it agrees well with the prediction.

7. Conclusions

The analysis contained in the paper is not complete in the sense that the designs of perforations and of the external shape of the structure should be conducted in parallel. In consequence the cells would no longer be square. Let us note however that formulae for substitute coefficients may be applied for such irregular cells as well. From variational interpretation given by Propositions $8 \div 11$ it follows that those coefficients will approximate the behaviour of the cells independently of their shapes. Hence the techniques used for perforation design are applicable also in general optimization, where both shape and perforation are taken into account.

The perforated material can be treated as a special example of a composite. Therefore this paper may be considered as a preliminary step toward a truly optimal design of structures. Combining methods of Sections 4 and 6 will be the subject of further study.

References

[1] N.W. Banichuk: Shape Optimization for Elastic Solids, Nauka, Moscow 1980 (in Russian).

[2] D. Chenais: On the existence of a solution in a domain identification problem, J. Math. Anal. Appl. 52 (1975), 189-219.

[3] P. Ciarlet: The Finite Element Method for Elliptic Problems, North-Holland, Amsterdam 1978.

[4] D. Cioranescu, J. Saint Jean Paulin: Homogenization in open sets with holes, J. Math. Anal. Appl., 71 (1979), 590-607.

[5] I. Hlavacek, J. Necas: On inequalities of Korn's type II, Arch. Rational Mech. Anal. 36 (1970), 313-334.

[6] P. Holnicki, A. Żochowski: Numerical methods for forecasting and

control of air pollution propagation, Systems Research Institute Research Report, no. ZTS-15-7/84, Warsaw, 1984 (in Polish).

[7] A. Marrocco, O. Pironneau: Optimum design with Lagrangian finite elements - design of an electromagnet, Comput. Math. Appl. Mech. Engng. 15 (1978).

[8] J.A. Nitsche: On Korn's second inequality, R.A.I.R.O., Numer. Anal. 15 (1981), 237-248.

[9] O. Pironneau: Optimal Shape Design for Elliptic Systems, book in preparation, 1982.

[10] B. Pshenichnyi: Linearization Algorithm, Nauka, Moscow 1982, (in Russian).

[11] L.A. Rozin: Variational Formulation of Problems for Elastic Systems, Leningrad, 1978 (in Russian).

[12] E. Sanchez-Palencia: Non-Homogeneous Media and Vibration Theory, Lect. Notes in Physics, Springer, Berlin, 1980.

[13] J. Simon: Differentiation with respect to the domain in boundary value problems, Numer. Funct. Anal. Optimiz. (1980), 649-687.

[14] G. Strang, G.J. Fix: An Analysis of the Finite Element Method, Prentice-Hall, Englewood Cliffs, 1973.

[15] A. Żochowski, K. Mizukami: Minimum-weight design with displacement constraints in 2-dimensional elasticity, Comput. Structures, 17, (1983), 365-369.

[16] A. Żochowski, K. Mizukami: A comparison of BEM and FEM in minimum weight design, in: Boundary Element, proc. of 5-th Int. Conf., C.A. Brebbia et al. ed., Springer, Berlin, 1983.

[17] A. Żochowski, K. Mizukami: A stress versus compliance constraint in a minimum weight design, Comput. Structures 18 (1984), 9-13.

[18] A. Żochowski: A variational interpretation of the homogenization procedure, to appear in Computers and Structures.

[19] A. Żochowski: Shape optimization in 2-dimensional elasticity, in: Constructive Aspects of Optimization, K. Malanowski, K. Mizukami (eds.), Polish Scientific Publ., Warsaw, 1985.

[20] V.V. Żikov, S.M. Kozlov, O.A. Oleinik, Ha Tien Ngoan: Homogenization and G-convergence of differential operators, Uspiekhy Mat. Nauk, 34 (1980), 65-128.

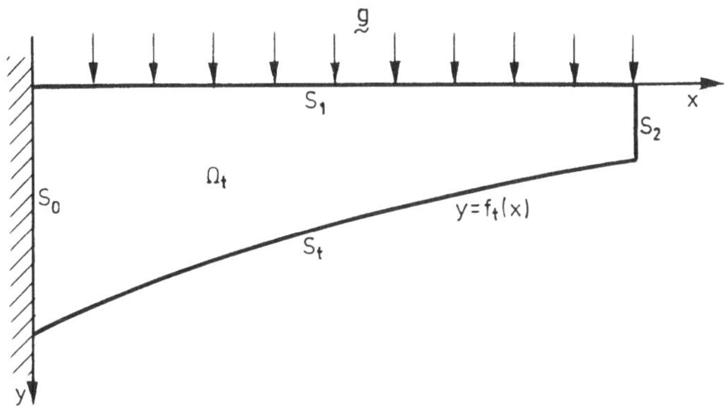

Fig. 1 Geometry of the problem.

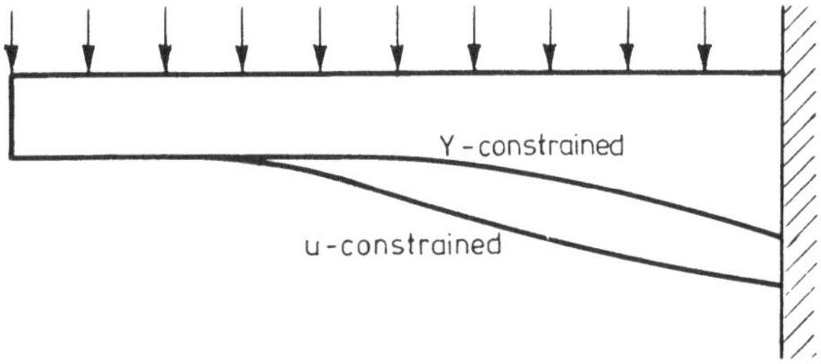

Fig. 2 Optimal desings of the beam.

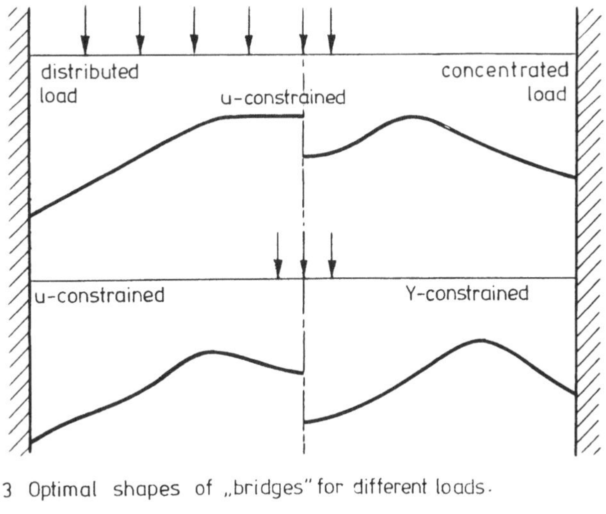

Fig. 3 Optimal shapes of „bridges" for different loads.

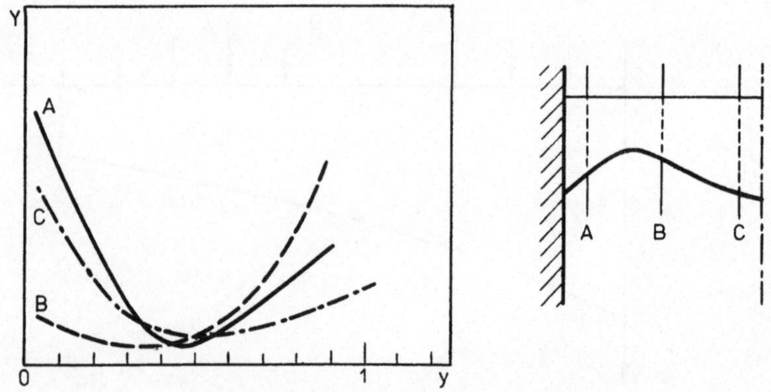

Fig. 4 Yield function distribution in the optimal shape.

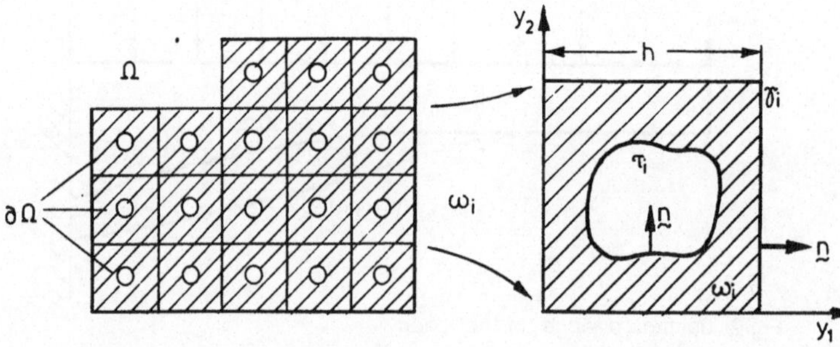

Fig. 5 The perforated domain

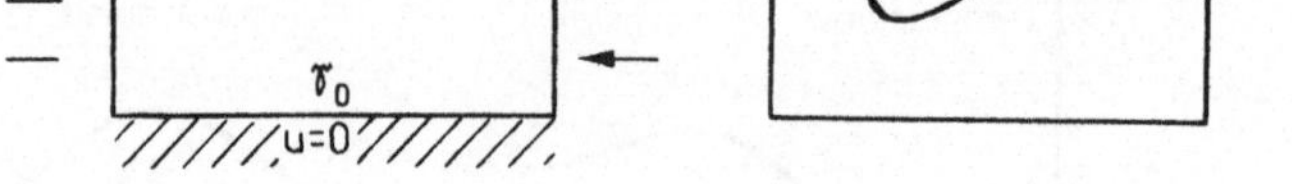

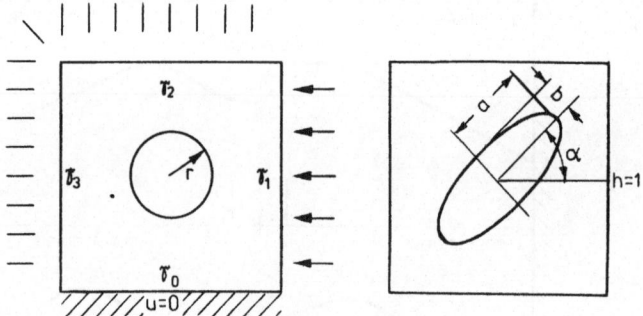

Fig. 6 The geometry of trial cells and test couditions,

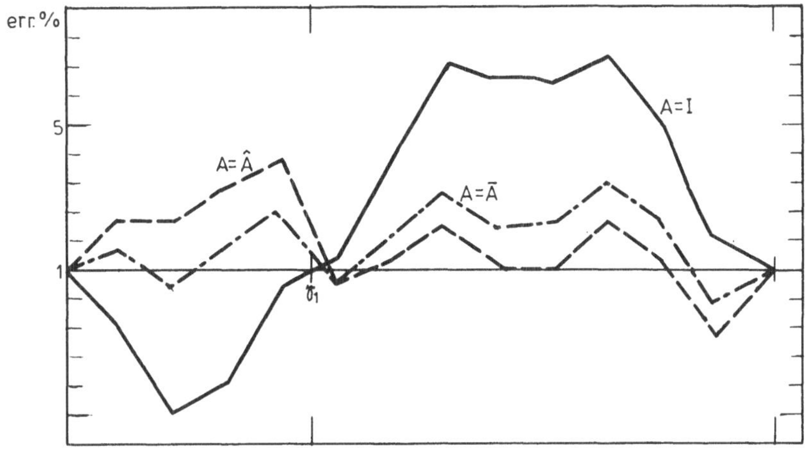

Fig. 7 The relative error for homogenized eq. 42, case I.

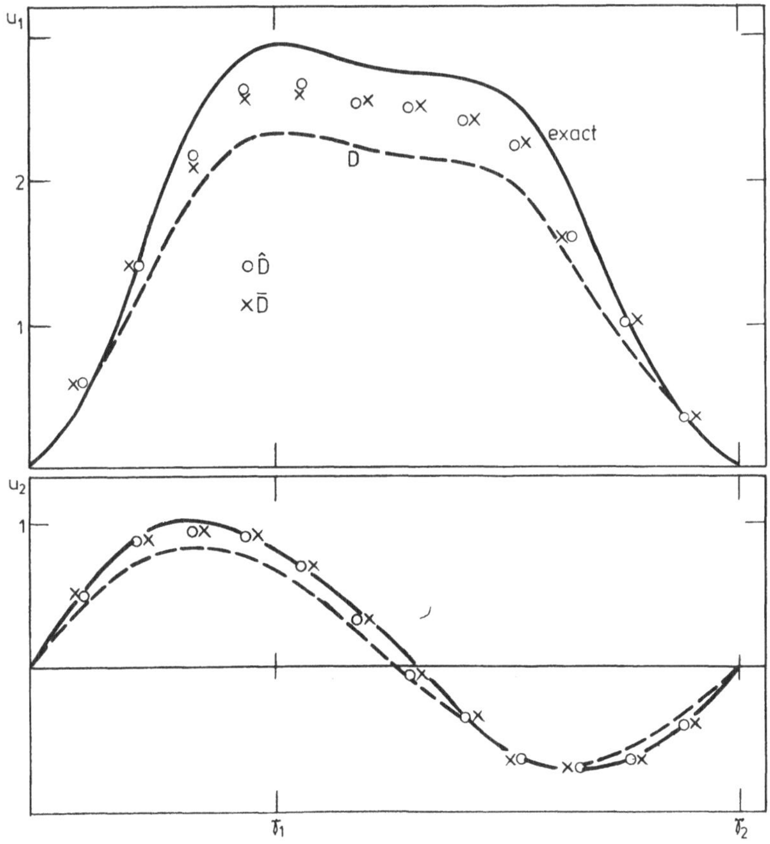

Fig. 8 A comparison of solutions for elactisity example.

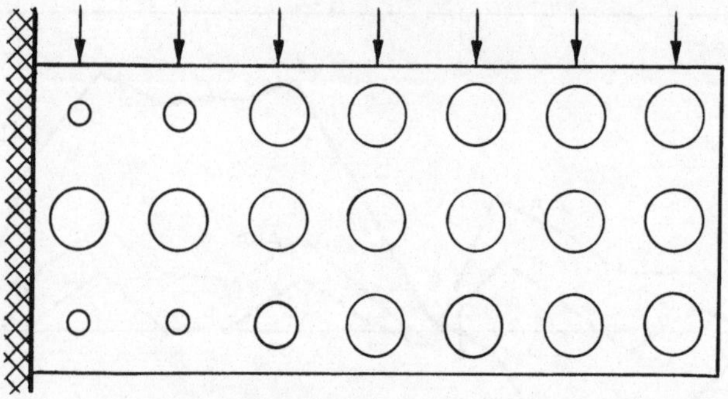

Fig.9 The beam with optimal perforation.

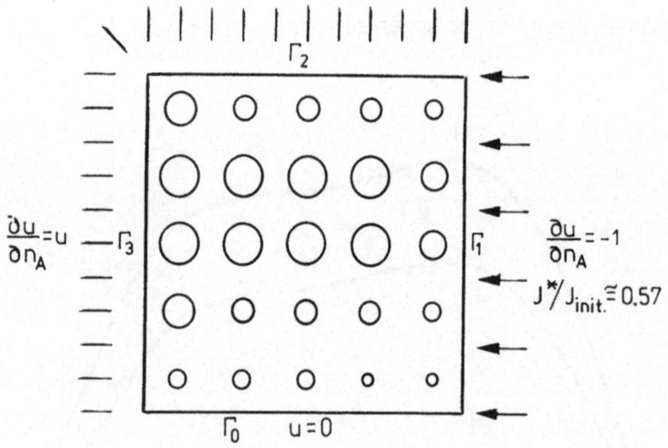

Fig.10 Optimal perforation - circular holes.

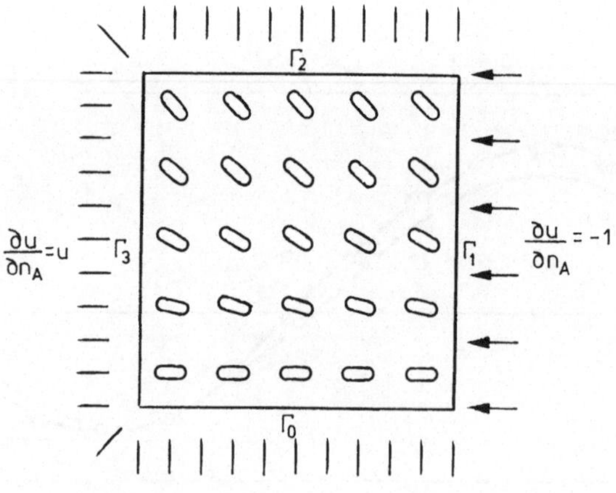

Fig.11 Optimal perforation - elliptical holes.

Chapter 6

NUMERICAL TREATMENT OF VARIATIONAL INEQUALITY GOVERNING MULTIDIMENSIONAL TWO-PHASE STEFAN PROBLEM

Irena Pawlow, Yuji Shindo and Yoshiyuki Sakawa

1. Introduction

Parabolic free boundary problems of the Stefan type arise as mathematical models of various heat flow and diffusion processes combined with phase transitions. In numerous applications, a prediction of the dynamical behaviour of such processes and their control (optimal control) are of interest [27,31] . A characteristic feature of the phase transitions of our concern manifests in the presence of some sets of singular points (in the form of some manifolds or so-called mushy regions) where parameters of the relevant model are discontinuous. Those sets belong to unknown variables of the process and are referred to as free boundaries. Their motion is characterized by appropriate physical balance laws. Despite of the primary interest in recovering the motion of the free boundaries, just this information usually appears rather short. In the multidimensional case, as a rule there are no results on the regularity of the free boundary, with exception for symmetric geometric configurations.

The classical solutions of the multidimensional Stefan problems are known only locally in time [14,20] . Results on the global in time existence of the solutions in the multidimensional situation are available only for variational (weak) formulations (the name fixed domain formulations is also used) in which the free boundary is no more explicitly present. However, the resulting weak solutions do not carry enough information to imply any regularity of the virtual free boundaries which are recovered as relevant level sets of those solutions. Despite of that deficiency, the variational approach appears to be an efficient way of the constructive treatment of the multidimensional problems of Stefan type.

Our main interest is concerned with setting up effective computational algorithms for solving multidimensional multiphase problems of the Stefan type, according to hints following from a numerical analysis of the problems which has been carried on in [26,28] . The numerical method developed in the present paper exploits the fixed domain formulation of the problems as a variational inequality (of the parabolic or mixed, parabolic-elliptic type). That inequality arises from expressing the Stefan problem in terms of the freezing index as a new dependent variable.

A theoretically justified approximation scheme corresponding to the derived variational inequality can be constructed by using first order finite elements in space and finite differences in time. Two variants of such schemes are introduced. The first of them prior to the discretization employs a regularization of the variational inequality whereas the second is based on the direct discretization.

Both schemes provide a simple time-stepping algorithm in which a nonlinear optimization problem is to be solved at each discrete time instant. More precisely, this problem consists in minimizing a convex, lower semicontinuous and radially unbounded functional over a finite dimensional space. This functional is Gateaux differentiable in the case of the regularized scheme and only Lipschitz continuous for the direct scheme.

We would like to mention that several techniques are applicable for solving the Stefan problems. Among the group of the fixed domain methods those based on the enthalpy formulation of the problem were studied most extensively [5,6,10,11,15-17,21-23,31-33]. The freezing index approach to the numerical solution of two-phase Stefan problems was applied in [1,2,11,13,18,24,26-29].

The presented approximation method is theoretically justified, i.e. results on the stability and convergence of the discrete solutions are available (cf., [28]). Our approach can be also applied to more general problems of the Stefan type, admitting quasilinear equations and nonlinear boundary conditions as well as some degenerations (cf., Remarks 3.2 and 3.3). The method is easily implementable, giving rise to a simple numerical algorithm. In a number of performed computational experiments, the algorithm has appeared efficient not only for the parabolic problems we consider (cf., also [29]) but also for degenerate elliptic-parabolic problems (cf., [24] and Example 4.2).

The present paper is an extension of our former publication [29].

2. Two-phase Stefan problem. Variational inequality formulation

2.1. Classical formulation

Let $\Omega \subset R^n$, $n > 1$, be an open bounded domain with a regular boundary Γ which consists of two disjoint parts Γ' and Γ''. At every time moment $t \in (0,T)$, where T is a positive and finite constant, Ω is decomposed into two subdomains $\Omega_1(t)$ and $\Omega_2(t)$ by a hypersurface $S(t)$ (representing the free boundary); then $\Omega = \Omega_1(t) \cup S(t) \cup \Omega_2(t)$. We shall denote $Q = \Omega\times(0,T)$, $Q_i = \cup_{t\in(0,T)} \Omega_i(t)\times\{t\}$, $i=1,2$; $S = \cup_{t\in(0,T)} S(t)\times\{t\}$, $\Sigma = \Gamma\times(0,T)$, $\Sigma' = \Gamma'\times(0,T)$, $\Sigma'' = \Gamma''\times(0,T)$. We have $Q \equiv Q_1 \cup S \cup Q_2$, $\Sigma = \Sigma' \cup \Sigma''$. $\vec{\nu}$ will denote the unit outward normal vector to Γ , $\vec{N}$ the unit vector normal to S , pointing Q_1 ; $\vec{N} \equiv (\vec{N}_x, N_t)$ where $\vec{N}_x$ is the projection of $\vec{N}$ onto R^n .

To fix some terminology, let us consider a typical Stefan problem in the context of the heat conduction combined with a phase transition, for instance, melting or solidification. Assume that in the domain Ω occupied by a heat conducting medium a subdomain $\Omega_1(t)$ corresponds to the solid and $\Omega_2(t)$ to the liquid phase. The phase transition is assumed to occur at temperature zero. In each phase, the governing equation which reflects the Fourier heat conduction is

$$c(\theta)\,\theta' - \nabla(k(\theta)\nabla\theta) = \begin{cases} \lambda_1 & \text{in } Q_1 \\ \lambda_2 & \text{in } Q_2 , \end{cases} \tag{2.1}$$

where $\theta = \theta(x,t)$ represents temperature, $c = c(\theta) > 0$ is the specific heat, $k = k(\theta) > 0$ is the thermal conductivity, and the density is assumed to be constant equal to unit; $\theta' \equiv \partial\theta/\partial t$. Because of different physical properties of the solid and liquid phases, the coefficients c and k are in general discontinuous at the phase transition temperature $\theta = 0$. The functions $\lambda_i = \lambda_i(x,t)$, $i=1,2$, represent internal heat sources in the solid and liquid phases, respectively. At the phase transition interface the following conditions are satisfied:

$$\theta = 0 \quad \text{on } S , \tag{2.2}$$

$$[k(\theta)\nabla\theta|_2 - k(\theta)\nabla\theta|_1] \cdot \vec{N}_x = L\,N_t \quad \text{on } S , \tag{2.3}$$

where $k(\theta)\nabla\theta|_i$, $i=1,2$, denote the unilateral limits of $k(\theta)\nabla\theta$ on S when approached from Q_i , L is a nonnegative constant representing the latent heat of the phase transition. The condition (2.2) expresses a local thermodynamic compatibility on the phase transition interface S and (2.3) represents the energy balance at S . Besides, the initial temperature and initial position of the interface S are

$$\theta(x,0) = \theta_o(x) \quad \text{for } x \in \Omega , \tag{2.4}$$

$$S(0) = S_o . \tag{2.5}$$

Appropriate boundary conditions on Σ are also to be imposed. We shall assume, for instance,

$$\theta = \bar{g}_1 \quad \text{on } \Sigma' , \tag{2.6}$$

$$k(\theta)\,\theta_\nu + p \int_0^\theta k(\xi)\,d\xi = u \quad \text{on } \Sigma'' , \tag{2.7}$$

where θ_ν denotes the outward normal derivative on Γ, $\bar{g}_1 = \bar{g}_1(x,t)$, $p = p(x) > 0$, $u = u(x,t)$ are given functions. The condition (2.6) prescribes temperature on Σ' and (2.7) the heat flux across Σ'' according to the Newton's law, with the coefficient $p(x)$ corresponding to the heat permeability of the boundary Γ''.

The above-formulated two-phase Stefan problem consists in determining a function $\theta : Q \to R$ and a surface $S \subset Q$ which satisfy the system (2.1)-(2.7). By a classical solution of the Stefan problem we mean the pair $\{\theta, S\}$ such that $\theta, \nabla\theta \in C^0(\bar{Q}_i)$, $\Delta\theta, \theta' \in C^0(Q_i)$, $i=1,2$, and S is an n-dimensional C^1-manifold.

Let us recall once more that on the contrary to the case $n = 1$ when numerous results on the global in time existence, uniqueness, stability and asymptotic behaviour of the solutions were available (cf., in particular [12,30,33]), in the multidimensional situation the only results on the existence of the classical solutions were local in time [14,20] .

The problem (2.1)-(2.7) with nonnegative coefficient c (possibly vanishing on a subdomain) arises, in particular, as a model of saturated-unsaturated flows in porous media [15] and electrochemical machining processes [11] . In the sequel, we shall refer to (2.1)-(2.7) as to the parabolic Stefan problem provided that c is strictly positive and as to the degenerate elliptic-parabolic Stefan problem if c is only nonnegative.

2.2. Variational formulation

In order to ensure the global in time existence of a solution to the problem (2.1)-(2.7), one usually turns to an appropriate variational formulation. The method we develop in this paper is based on a variational inequality representation. To get at it, we suitably transform the problem (2.1)-(2.7). At first, the so-called Kirchhoff transformation

$$\Theta = K(\theta) \triangleq \int_0^\theta k(\xi)\,d\xi \tag{2.8}$$

is applied. In terms of Θ , the problem (2.1)-(2.7) takes on the form

$$\begin{cases} \rho(\Theta)\,\Theta' - \Delta\Theta = \begin{cases} \lambda & \text{in } Q_1 \\ \lambda + \tilde{\lambda} & \text{in } Q_2 \;, \end{cases} & (2.9) \\ \Theta = 0 \quad \text{on } S \;, & (2.10) \\ [\nabla\Theta|_2 - \nabla\Theta|_1] \cdot \vec{N}_x = L\, N_t \quad \text{on } S \;, & (2.11) \\ \Theta(0) = \Theta_o \;, \quad S(0) = S_o \quad \text{in } \Omega \;, & (2.12) \\ \Theta = g_1 \quad \text{on } \Sigma' \;, & (2.13) \\ \Theta_\nu + p\,\Theta = u \quad \text{on } \Sigma'' \;, & (2.14) \end{cases}$$

where

$$\lambda = \lambda_1 \;, \quad \tilde{\lambda} = \lambda_2 - \lambda_1 \;, \quad \rho(\Theta) = \frac{c(K^{-1}(\Theta))}{k(K^{-1}(\Theta))} \;,$$

$$g_1 = K(\bar{g}_1) \;, \quad \Theta_o = K(\theta_o) \;,$$

K^{-1} denotes the inverse of K which is well-defined due to the postulated strict positivity of the coefficient k .

Now we transform the problem (2.9)-(2.14) (for the time-being in a formal way) and give the resulting formulations. Observe that by employing the characteristic function of the set Q_2 in Q ,

$$\chi = \begin{cases} 1 & \text{in } Q_2 \\ 0 & \text{in } Q \setminus Q_2 \;, \end{cases} \tag{2.15}$$

the internal part (2.9)-(2.11) of the problem may be inserted into the equation

$$\rho(\Theta)\,\Theta' - \Delta\Theta = \lambda + \tilde{\lambda}\,\chi - L\,\chi' \;, \tag{2.16}$$

to be satisfied in the space of distributions $\mathcal{D}'(Q)$ (cf., in particular [19]). Introduce the multivalued function

$$\gamma_o(r) \triangleq \tilde{\gamma}(r) + L\,\mathrm{sign}^+(r) \;, \quad r \in \mathbb{R} \;, \tag{2.17}$$

representing the enthalpy, where

$$\tilde{\gamma}(r) = \int_0^r \rho(\xi)\, d\xi$$

and $\mathrm{sign}^+(r)$ is the Heaviside's graph

$$\mathrm{sign}^+(r) \triangleq \begin{cases} 0 & \text{if } r < 0 \\ [0,1] & \text{if } r = 0 \\ 1 & \text{if } r > 0 \;. \end{cases}$$

Note that obviously $\chi \in \mathrm{sign}^+(\Theta)$ a.e. in Q. Hence, in terms of a measurable selection w of the graph γ_o,

$$w \in \gamma_o(\Theta) \quad \text{a.e. in } Q, \tag{2.18}$$

the equation (2.16) may be rewritten as

$$w' - \Delta\Theta = \lambda + \tilde{\lambda}\chi \quad \text{in } \mathcal{D}'(Q). \tag{2.19}$$

The initial condition proper for (2.19) is

$$w(0) = w_o \quad \text{in } \Omega. \tag{2.20}$$

To provide an internal compatibility of the problem statement, the initial enthalpy w_o should be adjusted to the initial temperature Θ_o so that

$$w_o \in \gamma_o(\Theta_o) \quad \text{a.e. in } \Omega. \tag{2.21}$$

In the parabolic case, Θ_o is uniquely determined by (2.21) because the inverse γ_o^{-1} is then a singlevalued function. In the degenerate situation when ρ is only nonnegative, one needs both data. To complete the system, the boundary conditions (2.13),(2.14) should be added to (2.18)-(2.21).

In this way, the two-phase Stefan problem (2.1)-(2.7) has been transformed to the form (2.18)-(2.21),(2.13),(2.14) which is usually referred to as the enthalpy fixed domain formulation. This formulation gives rise to various notions of weak solutions and the corresponding approximation schemes.

An alternative approach is based on a reformulation of the Stefan problem as a variational inequality [8,13]. To this purpose, introduce the new dependent variable

$$\bar{y}(x,t) \triangleq \int_0^t \Theta(x,\tau)\, d\tau \quad, \quad (x,t) \in Q, \tag{2.22}$$

called freezing index. Formally, the problem (2.18)-(2.21),(2.13),(2.14) expressed in terms of $\bar{y}$ takes on the form

$$\begin{cases} \gamma_o(\bar{y}') - X_o(\bar{y}') - \Delta\bar{y} \ni f_o & \text{in } Q, \quad (2.23) \\ \bar{y}(0) = 0 & \text{in } \Omega, \quad (2.24) \\ \bar{y} = G_1 & \text{on } \Sigma', \quad (2.25) \\ \bar{y}_\nu + p\,\bar{y} = G_2 & \text{on } \Sigma'', \quad (2.26) \end{cases}$$

where

$$[X_o(\bar{y}')](x,t) = {}_0\int^t \tilde{\lambda}(x,\tau)\,\chi(x,\tau)\,d\tau \quad , \quad \chi \in \mathrm{sign}^+(\bar{y}') \quad ,$$

$$f_o(x,t) = \Lambda(x,t) + w_o(x) \quad , \quad \Lambda(x,t) = {}_0\int^t \lambda(x,\tau)\,d\tau \quad ,$$

$$G_1(x,t) = {}_0\int^t g_1(x,\tau)\,d\tau \quad , \quad G_2(x,t) = {}_0\int^t u(x,\tau)\,d\tau \quad .$$

The problem (2.23)-(2.26) may be formulated as an evolution variational inequality of the second kind (according to the terminology of [9]). We shall use the following notations:

$H = L^2(\Omega)$, $V = H^1(\Omega)$ appropriately with the standard norms $\|\cdot\|_H$, $\|\cdot\|_V$; $(\cdot,\cdot)$, $(\cdot,\cdot)_{\Gamma''}$ are the scalar products in H and $L^2(\Gamma'')$, respectively;

$K(t) = \{ z \in V \mid z|_{\Gamma'} = g_1(t) \}$, $V_o = \{ z \in V \mid z|_{\Gamma'} = 0 \}$.

Note that V_o is a closed subspace of V and can be equipped with the norm

$$\|z\|^2_{V_o} = (\nabla z,\nabla z) + (p\,z,z)_{\Gamma''} \tag{2.27}$$

which is equivalent over V_o to the standard norm of V , provided

$p \in L^\infty(\Gamma'')$, $p > 0$ and the Lebesgue measure of the set $\Gamma' \cup \{ x \in \Gamma'' \mid p(x) > 0 \}$ is positive. (2.28)

Introduce also the bilinear form

$$a(y,z) = (\nabla y,\nabla z) + (p\,y,z)_{\Gamma''} \tag{2.29}$$

on $V\times V$, and the functional $\Psi_o : V \to R$ defined by

$$\Psi_o(z) \triangleq L \int_\Omega \psi_o(z(x))\,dx \qquad \text{where} \quad \psi_o(r) \triangleq \max\{0,r\} \ . \tag{2.30}$$

According to the definition (2.17) of γ_o , the equation (2.23) can be given the form

$$-\tilde{\gamma}(\bar{y}') + X_o(\bar{y}') + \Delta\bar{y} + f_o \in L\,\partial\psi_o(\bar{y}') \quad \text{in } Q \ , \tag{2.31}$$

where $\partial\psi_o(r)$ is the subdifferential of the function ψ_o at the point r . By the definition of the subdifferential of a convex function, the system (2.31),(2.24)-(2.26) gives rise to the following formulation:

Determine a function $\bar{y} : [0,T] \to V$ such that

$\bar{y}(t) \in V$, $\bar{y}'(t) \in K(t)$ for a.a. $t \in [0,T]$,

as well as

$$\begin{cases} (\tilde{\gamma}(\bar{y}'(t))-[X_o(\bar{y}')](t) - f_o(t) \ , \ z-\bar{y}'(t)\ ) + a(\bar{y}(t),z-\bar{y}'(t)) - \\ \quad - (G_2(t),z-\bar{y}'(t))_{\Gamma''} + \Psi_o(z) - \Psi_o(\bar{y}'(t)) \geq 0 \\ \quad \text{for all} \quad z \in K(t) \ , \quad \text{a.a.} \quad t \in [0,T] \ , \\ \text{where} \quad \chi \quad \text{is measurable in} \quad Q \ , \quad \chi \in \text{sign}^+(\bar{y}') \quad \text{a.e. in} \quad Q \ , \\ \bar{y}(0) = 0 \quad \text{in} \quad \Omega \ . \end{cases} \tag{2.32}$$

For further convenience, we transform the system (2.32) so that to get the homogeneous boundary condition on Γ'. To this end, we introduce the shifted variable

$$y = \bar{y} - G \quad \text{in} \quad Q \ , \tag{2.33}$$

where $G(x,t) \triangleq {}_0\int^t g(x,\tau)\, d\tau$ and $g \in L^\infty(0,T;V)$ is an extension of g_1 onto $\bar{Q}$, such that $g|_{\Gamma'} = g_1$. In terms of y, (2.32) can be equivalently formulated as

$$\begin{cases} \underline{\text{Problem (VI)}}: \quad \text{Determine a function} \quad y : [0,T] \to V_o \quad \text{such} \\ \text{that} \quad y\ ,\ y' \in L^2(0,T;V_o) \ , \\ (\tilde{\gamma}(y'(t)+G'(t))-[X_o(y'+G')](t)-f_o(t) \ , \ z-y'(t)\ ) + \\ \quad + a(y(t)+G(t),z-y'(t)) - (G_2(t),z-y'(t))_{\Gamma''} + \\ \quad + \Psi_o(z+G'(t)) - \Psi_o(y'(t)+G'(t)) \geq 0 \\ \quad \text{for all} \quad z \in V_o \ , \quad \text{a.a.} \quad t \in [0,T] \ , \\ \text{where} \quad \chi \quad \text{is measurable in} \quad Q \ , \quad \chi \in \text{sign}^+(y'+G') \quad \text{a.e. in} \quad Q \ , \\ y(0) = 0 \quad \text{in} \quad \Omega \ . \end{cases} \tag{2.34}$$

<u>Definition</u>. By the weak solution of the Stefan problem (2.1)-(2.7) we shall mean a function y which satisfies (VI) .

Clearly, any classical solution of (2.1)-(2.7) satisfies (VI) and, conversely, $\theta = K^{-1}(y'+G')$ would fulfil (2.1)-(2.7) in the classical sense, provided that y and G are sufficiently regular.

According to the above definition, the free boundary S may be recovered only à posteriori as the level set

$$S = \{ (x,t) \in Q \mid y'(x,t) + G'(x,t) = 0 \} \ . \tag{2.35}$$

This characterization, to become more specific, would require some higher regularity of y' than the continuity which is available in the multidimensional case [3,7] . Lacking this property, one should interpret (2.35) as a condition that eliminates those subdomains of Q which cannot contain the free boundary rather than as an accurate specification of S .

2.3. Existence and uniqueness of the solution to (VI)

The following hypotheses are maintained throughout the paper:

(A1) γ_o is defined by (2.17) , where L is a nonnegative constant and ρ is a given function, possibly discontinuous at $r = 0$ and globally bounded:

$$0 < \bar{\rho} \leq \rho(r) \leq \bar{\bar{\rho}} < +\infty \qquad \text{for } r \in R ;$$

(A2) $\lambda \in L^2(Q)$, $\tilde{\lambda} \in L^2(Q)$;

(A3) $g \in W^{1,\infty}(0,T;V) \cap H^2(0,T;H)$, $g|_{\Gamma'} = g_1$;

(A4) $u \in H^1(0,T;L^2(\Gamma''))$;

(A5) $\Theta_o \in V \cap L^\infty(\Omega)$ and $w_o = \gamma^0(\Theta_o)$ in Ω , where γ^0 denotes the minimal norm section of the graph γ , uniquely defined due to the maximal monotonicity of γ ;

(A6) the condition (2.28) is satisfied.

Note that the above hypotheses bring about the following consequences. By (A1), the mapping $\tilde{\gamma} : H \to H$ is Lipschitz continuous:

$$\|\tilde{\gamma}(y) - \tilde{\gamma}(z)\|_H \leq \bar{\bar{\rho}} \|y - z\|_H \quad \text{for all } y,z \in H , \tag{2.36}$$

and strictly monotone:

$$(\tilde{\gamma}(y)-\tilde{\gamma}(z) , y-z) \geq \bar{\rho} \|y - z\|_H^2 \quad \text{for all } y,z \in H . \tag{2.37}$$

The functional $\Psi_o : V \to R$ is bounded, convex and lower semicontinuous. The bilinear form $a(\cdot,\cdot)$ is symmetric, continuous:

$$|a(y,z)| \leq \bar{\bar{\omega}} \|y\|_V \|z\|_V \quad \text{for all } y,z \in V ; \quad \bar{\bar{\omega}} > 0 \tag{2.38}$$

and, due to (A6), V-elliptic:

$$a(y,y) \geq \bar{\omega} \|y\|_V^2 \quad \text{for all } y \in V , \quad \bar{\omega} > 0 . \tag{2.39}$$

Under the hypotheses (A1)-(A6), there exists at least one solution y of (VI) , such that

$$y \in W^{1,\infty}(0,T;V) \cap H^2(0,T;H) , \quad y'(0) = \Theta_o , \tag{2.40}$$

and the estimates

$$\|y\|_{W^{1,\infty}(0,T;V)} + \|y''\|_{L^2(Q)} \leq C \tag{2.41}$$

hold with a constant C dependent only upon the bounds on the data (cf., [25]).

Provided the assumption

(A7) $\tilde{\lambda} = 0$ a.e. in Q ,

the solution y of (VI) can be shown to be unique (cf., [25]).

Remark 2.1. The hypothesis (A1) may be relaxed by admitting the degeneration of parabolicity related to $\bar{\rho} = 0$. In that case, the assumption (A2) is to be replaced by a stronger one:

$\lambda \in H^1(0,T;H)$.

This suffices to ensure the existence and uniqueness of the solution Θ , à priori bounded in $W^{1,\infty}(0,T;V)$ (cf., [25]).

3. Construction of discrete approximations

3.1. Regularized problem

Proceeding in a standard way, we shall introduce a family of regularized problems corresponding to (VI). The regularization is understood here in the sense of approaching (VI) by a family of variational equations. Let $\Psi_\varepsilon : V \to R$, $\varepsilon > 0$, be defined by

$$\Psi_\varepsilon(z) \triangleq L \int_\Omega \psi_\varepsilon(z(x))\, dx , \tag{3.1}$$

where ψ_ε is a smooth approximation to ψ_0 which in particular can be taken in the form

$$\psi_\varepsilon(r) \triangleq \begin{cases} 0 & \text{if } r < 0 \\ \dfrac{r^3}{\varepsilon^2}\left(1 - \dfrac{r}{2\varepsilon}\right) & \text{if } 0 < r < \varepsilon \\ r - \dfrac{\varepsilon}{2} & \text{if } r > \varepsilon \end{cases} . \tag{3.2}$$

One can see that $\psi_\varepsilon \in C^2(R)$ and it approximates ψ_0 with

$$| \psi_0(r) - \psi_\varepsilon(r) | < \frac{\varepsilon}{2} \quad \text{for all } r \in R . \tag{3.3}$$

The above approximation of ψ_0 induces the following compatible approximations to the terms X_0 and f_0 :

$$[X_\varepsilon(y)](x,t) = \int_0^t \tilde{\lambda}(x,\tau)\, D\psi_\varepsilon(y(x,\tau))\, d\tau ,$$

$$f_\varepsilon(x,t) = \Lambda(x,t) + w_\varepsilon(x) , \quad w_\varepsilon(x) = \tilde{\gamma}(\Theta_0(x)) + L\, D\psi_\varepsilon(\Theta_0(x)), \tag{3.4}$$

where the notation $D\psi_\varepsilon(r) = \dfrac{d\psi_\varepsilon(r)}{dr}$ has been used.

The regularized problem corresponding to (VI) takes on then the form:

$$\begin{cases} \underline{\text{Problem } (VI)_\varepsilon} : \text{ Determine a function } y_\varepsilon : [0,T] \to V_o , \\ \text{such that } y_\varepsilon , y'_\varepsilon \in L^2(0,T;V_o) , \\ (\tilde{\gamma}(y'_\varepsilon(t)+G'(t))-[X_\varepsilon(y'_\varepsilon+G')](t)-f_\varepsilon(t) , z-y'_\varepsilon(t)) + \\ \quad + a(y_\varepsilon(t)+G(t),z-y'_\varepsilon(t)) - (G_2(t),z-y'_\varepsilon(t))_{\Gamma''} + \\ \quad + \Psi_\varepsilon(z+G'(t)) - \Psi_\varepsilon(y'_\varepsilon(t)+G'(t)) \geq 0 \\ \quad \text{for all } z \in V_o , \text{ a.a. } t \in [0,T] , \\ y_\varepsilon(0) = 0 \quad \text{in } \Omega . \end{cases} \tag{3.5}$$

Due to the differentiability of Ψ_ε, the inequality (3.5) is equivalent to the equation

$$(\gamma_\varepsilon(y'_\varepsilon(t)+G'(t))-[X_\varepsilon(y'_\varepsilon+G')](t)-f_\varepsilon(t) , z) + a(y_\varepsilon(t)+G(t),z) - (G_2(t),z)_{\Gamma''} = 0 \quad \text{for all } z \in V_o , \text{ a.a. } t \in [0,T] , \tag{3.5'}$$

where $\gamma_\varepsilon(r) = \tilde{\gamma}(r) + L\, D\psi_\varepsilon(r) , \quad r \in R$.

Notice that formally the regularized problem corresponding to (2.23)-(2.26) takes on the the form

$$\begin{cases} \gamma_\varepsilon(\bar{y}'_\varepsilon) - X_\varepsilon(\bar{y}'_\varepsilon) - \Delta\bar{y}_\varepsilon = f_\varepsilon \quad \text{in } Q , \\ (2.24)-(2.26) , \end{cases} \tag{3.6}$$

where $\bar{y}_\varepsilon = y_\varepsilon + G$. For further purposes, it is useful to observe that by (3.2) we have

$$0 \leq D\psi_\varepsilon(r) \leq 1 , \quad 0 \leq D^2\psi_\varepsilon(r) \leq \frac{3}{2\varepsilon} \quad \text{for } r \in R$$

and, therefore, according to (A1) ,

$$|D\gamma_\varepsilon(r)| \leq \frac{C}{\varepsilon} \quad \text{for } r \in R \tag{3.7}$$

with a constant C independent of ε.

We recall that, as for (VI) , under the hypotheses (A1)-(A6) there exists a solution y_ε of $(VI)_\varepsilon$ which satisfies the bounds (2.41) uniformly with respect to ε. Moreover, y_ε can be shown to satisfy the bound

$$\|y'_\varepsilon\|_{L^2(0,T;H^2(\Omega))} \leq \frac{C}{\varepsilon^{1/2}} \tag{3.8}$$

with a constant C independent of ε (cf., [28]). This bound appears crucial for establishing error estimates for discrete approximations (cf., Section 3.4). Provided that (A7) holds, the solution y_ε of $(VI)_\varepsilon$ is unique.

In the sequel, we shall identify (VI) and $(VI)_\varepsilon$ with $\varepsilon = 0$.

Remark 3.1. In the degenerate case, as admitted in Section 2, the whole consideration is to be preceded by a parabolic regularization of (VI) (cf., [28], for instance). To this end, we introduce an auxiliary family of strictly monotone functions

$$\tilde{\gamma}_\mu(r) \triangleq \tilde{\gamma}(r) + \mu r, \qquad r \in R$$

with $\mu > 0$. As it has been proved in [28], the difference between the unique solutions of the regularized and original problems admits the estimate

$$\| y_\mu - y \|_{L^\infty(0,T;V)} + \mu^{1/2} \| y'_\mu - y' \|_{L^2(Q)} \leq C \mu^{1/2}$$

with a constant C independent of μ.

3.2. Discrete approximations

To simplify our further exposition, let us assume that

(A8) Ω is a convex polygonal domain in R^2;

(A9) $\Theta_o \in H^2(\Omega)$.

Let T_h denote a triangulation of Ω, with the mesh parameter $h \in (0,1]$. The triangulation is assumed to be regular [4].

Denote by $I_{p_o} = \{1,\ldots,p_o\}$ the set of indices j associated with the internal nodes x_j of the triangulation, $I_p = \{1,\ldots,p\}$ the set of indices corresponding to the nodes in the interior of Ω and on Γ'', and $I_q = \{1,\ldots,q\}$ the set containing the indices of all nodes. $\tilde{V}_h$ will denote a finite dimensional subspace of V, defined by

$$\tilde{V}_h \triangleq \{ v_h \in V \cap C^0(\bar{\Omega}) \mid v_h \text{ is an affine function over each } T \in T_h \}$$

and $V_h \triangleq \{ v_h \in \tilde{V}_h \mid v_h|_{\Gamma'} = 0 \}$.

We introduce a basis $\{w_j^h\}_{j \in I_q}$ of the space $\tilde{V}_h$, with the functions w_j^h defined by:

$$w_j^h \in \tilde{V}_h, \quad w_j^h(x_j) = 1, \quad w_j^h(x_k) = 0 \text{ for } k \neq j .$$

Then $\tilde{V}_h$ admits the representation $\tilde{V}_h = \operatorname{span}_{j \in I_q} \{w_j^h\}$ and any function $v \in C^0(\bar{\Omega})$ can be uniquely interpolated in $\tilde{V}_h$ by

$$I_h v(x) = \sum_{j \in I_q} v(x_j)\, w_j^h(x) .$$

By the postulated properties of the triangulation T_h, the space

$\tilde{V}_h$ satisfies [4] : (i) approximation property

$$\|v - I_h v\|_H + h \|v - I_h v\|_V \leq C h^2 \|v\|_{H^2(\Omega)} \quad \text{for all} \quad v \in H^2(\Omega) , \tag{3.9}$$

where C is a positive constant independent of h ;
(ii) discrete inverse norm inequality

$$\|v_h\|_V \leq S(h) \|v_h\|_H \quad \text{for all} \quad v_h \in \tilde{V}_h , \tag{3.10}$$

where $S(h) = C/h$, C is an independent of h finite constant which is specified by the concrete type of the triangulation.

The time interval $[0,T]$ will be divided into N equal subintervals $[t_i,t_{i+1}]$, $i=0,1,\dots,N-1$, where $t_i = ik$, $k = T/N$.

In the sequel, we shall use the standard notations:

$$y^i(x) = y(x,t_i) , \quad \delta y^i = \frac{y^{i+1}-y^i}{k} , \quad y^{i+\alpha} = y^i + \alpha k\, \delta y^i ,$$

$$\delta y^{i+\alpha} = \frac{y^{i+1+\alpha}-y^{i+\alpha}}{k} = \delta y^i + \alpha\, (\delta y^{i+1}-\delta y^i) , \quad \alpha \in [0,1] ,$$

$$\delta^2 y^i = \frac{\delta y^{i+1}-\delta y^i}{k} = \frac{y^{i+2} - 2y^{i+1} + y^i}{k^2} .$$

We introduce two types of discrete schemes corresponding to (VI) , one exploiting the regularization with a parameter $\varepsilon > 0$ and the other - direct $(\varepsilon = 0)$. The schemes are given the following joint formulation:

$$\left\{\begin{array}{l}
\underline{\text{Problem } (VI)_{\varepsilon,hk}} \quad (\varepsilon \geq 0 \text{ is any arbitrary parameter }) : \\
\text{Determine } y^i_{\varepsilon hk} \in V_h , \quad i=0,1,\dots,N , \text{ which satisfy} \\
y^{i+1}_{\varepsilon hk} = y^i_{\varepsilon hk} + k\, \delta y^i_{\varepsilon hk} , \\
(\tilde{\gamma}(\delta y^i_{\varepsilon hk}+\delta G^i_h) - [X_\varepsilon(\delta y_{\varepsilon hk}+\delta G_h)]^i - f^{i+\bar{\alpha}}_\varepsilon , z_h - \delta y^i_{\varepsilon hk}) + \\
\qquad + a(y^{i+\alpha}_{\varepsilon hk}+G^{i+\alpha}_h , z_h-\delta y^i_{\varepsilon hk}) - (G^{i+\bar{\alpha}}_2 , z_h-\delta y^i_{\varepsilon hk})_{\Gamma''} + \\
\qquad + \Psi_\varepsilon(z_h+\delta G^i_h) - \Psi_\varepsilon(\delta y^i_{\varepsilon hk}+\delta G^i_h) \geq 0 \\
\qquad \text{for all } z_h \in V_h , \quad i=0,1,\dots,N-1 , \\
y^0_{\varepsilon hk} = 0 \quad \text{in } \Omega ,
\end{array}\right. \tag{3.11}$$

where $\alpha, \bar{\alpha} \in [0,1]$ are arbitrary parameters,

$$[X_\varepsilon(y)]^i = \sum_{\nu=0}^{i-1} k\, \tilde{\lambda}^\nu\, D\psi_\varepsilon(y^\nu) \quad \text{if } \varepsilon > 0 , \quad \text{whereas at } \varepsilon = 0$$

$$[X_0(y)]^i = \sum_{\nu=0}^{i-1} k\, \tilde{\lambda}^\nu\, \chi^\nu , \quad \chi^\nu = \begin{cases} 0 & \text{if } y^\nu \leq 0 \\ 1 & \text{if } y^\nu > 0 . \end{cases}$$

The introduced approximating schemes can be implemented in the form of a time-stepping algorithm which gives rise to a numerical method for solving the problem (VI) :

Step 0. Select ε , h , k . Set $y^0_{\varepsilon hk} = 0$, $i=0$.

Step 1 . Solve (3.11) with respect to $\delta y^i_{\varepsilon hk}$.

Step 2 . Compute $y^{i+1}_{\varepsilon hk} = y^i_{\varepsilon hk} + k\,\delta y^i_{\varepsilon hk}$.

Step 3 . If $i = N-1$, then Stop. Otherwise, set $i := i+1$ and return to Step 1 .

To perform Step 1 of the algorithm, it is useful to note that for any fixed i , $\delta y^i_{\varepsilon hk}$ may be characterized as

$$\delta\bar{y}^i_{\varepsilon hk} = \underset{\substack{\bar{z}_h = z_h + \delta G^i_h \\ z_h \in V_h}}{\arg\inf}\; J^i_\varepsilon(\bar{z}_h) \;, \quad i=0,\dots,N-1 \;, \quad \delta\bar{y}^i_{\varepsilon hk} = \delta y^i_{\varepsilon hk} + \delta G^i_h \;, \tag{3.12}$$

where
$$J^i_\varepsilon(z) = \frac{1}{2}\alpha\, k\, a(z,z) + B(z) + \Psi_\varepsilon(z) + l^i_\varepsilon(z) \;, \tag{3.13}$$

$$B(z) = \int_\Omega \beta(z(x))\, dx \;, \qquad \beta(z) = \int_0^z \tilde{\gamma}(\xi)\, d\xi \;,$$

$$l^i_\varepsilon(z) = a(y^i_{\varepsilon hk} + G^i_h, z) - ([X_\varepsilon(\delta y_{\varepsilon hk} + \delta G_h)]^i + f^{i+\bar{\alpha}}_\varepsilon \;, z) - (G^{i+\bar{\alpha}}_2, z)_{\Gamma''} \;.$$

According to the above definition of $J^i_\varepsilon : V_h \to R$, the existence and uniqueness of the solution $y_{\varepsilon hk}$ to problem $(VI)_{\varepsilon,hk}$ can be deduced directly by the Weierstrass theorem. Indeed, notice that the term $\alpha\, k\, a(z,z)$ in (3.13) is convex, lower semicontinuous and, by (2.39),

$$\alpha\, k\, a(z,z) \geq \alpha k\, \bar{\omega}\, \|z\|^2_V \qquad \text{for } z \in V \;.$$

The functional $B : V_h \to R$ is continuous and strictly convex, since its Gateaux differential $DB(\cdot)$, characterized by

$$(DB(y), z) = (\tilde{\gamma}(y), z) \qquad \text{for } y,z \in H \;,$$

is strictly monotone due to (2.37) . In addition, by (A1),

$$B(z) \geq \frac{\bar{\rho}}{2}\|z\|^2_H \qquad \text{for all } z \in H \;.$$

Since $\Psi_\varepsilon : V_h \to R$ is convex and lower semicontinuous, we eventually conclude that the functional $J^i_\varepsilon : V_h \to R$ is strictly convex, lower semicontinuous and radially unbounded for each $\varepsilon > 0$, $\alpha \in [0,1]$ and any fixed $h,k > 0$.

<u>Proposition 3.1</u>. <u>For any</u> $\varepsilon > 0, \alpha, \bar{\alpha} \in [0,1]$, <u>the problem</u> $(VI)_{\varepsilon,hk}$ <u>has the unique solution</u> $y_{\varepsilon hk}$.

We remark that if $\varepsilon > 0$ then the functional $J^i_\varepsilon(\cdot)$ is Gateaux differentiable. This property plays an underlying role at numerical solving the minimization problem (3.12) .

The presented approximation schemes admit various extensions.

Remark 3.2. The schemes are also applicable in the case of Stefan problems with variable (explicitly space and time dependent) coefficients, nonlinear internal heat sources and nonlinear boundary flux terms (cf., [25,29]).

Remark 3.3. The area of applicability of our approach covers also the degenerate Stefan problems, with vanishing coefficient ρ (cf., [24,28]).

3.3. Nonlinear minimization problem

In order to completely characterize the method, it remains to specify numerical quadrature formulae. We shall consider the following finite dimensional problem corresponding to (3.12) :

$$\text{Determine } \delta\bar{Y}^i_{\varepsilon h} = \underset{\substack{\bar{z}_h = z_h + \delta G^i_h \\ z_h \in V_h}}{\arg\inf} J^i_{\varepsilon h}(\bar{z}_h) , \quad \delta\bar{Y}^i_{\varepsilon h} = \delta Y^i_{\varepsilon h} + \delta G^i_h , \tag{3.14}$$

where

$$J^i_{\varepsilon h}(z) = \frac{1}{2}\alpha k\, a_h(z,z) + B_h(z) + \Psi_{\varepsilon h}(z) + l^i_{\varepsilon h}(z) , \tag{3.15}$$

$$a_h(y,z) = (\nabla y, \nabla z)_h + (p\, y, z)_{\Gamma'',h} ,$$

$$(y,z)_h = \int_\Omega I_h[y(x)z(x)]\, dx \quad \text{for } y,z \in V \cap C^0(\bar{\Omega}) ,$$

$$(y,z)_{\Gamma'',h} = \int_{\Gamma''} I_h[y(x)z(x)]\, d\Gamma \quad \text{for } y,z \in V \cap C^0(\bar{\Omega}) ,$$

$$B_h(z) = \int_\Omega I_h[\beta(z(x))]\, dx , \quad \Psi_{\varepsilon h}(z) = L\int_\Omega I_h[\psi_\varepsilon(z(x))]\, dx ,$$

$$l^i_{\varepsilon h}(z) = a_h(\bar{Y}^i_{\varepsilon h}, z) - ([X_\varepsilon(\delta\bar{Y}_{\varepsilon h})]^i + f^{i+\bar{\alpha}}_\varepsilon , z)_h - (G^{i+\bar{\alpha}}_2, z)_{\Gamma'',h} .$$

Let $\bar{Z} = \{\bar{z}_1,\dots,\bar{z}_q\}$ and denote by $\langle\cdot,\cdot\rangle$ the scalar product in R^q. Since $\bar{z}_h = z_h + \delta G^i_h$ admits the representation $\bar{z}_h(x) = \sum_{j\in I_q} \bar{z}_j\, w^h_j(x)$, where

$$\bar{z}_j = z_j + \delta G^i_j \in R , \quad \delta G^i_h(x) = \sum_{j\in I_q} \delta G^i_j\, w^h_j(x) = I_h[\delta G^i] ,$$

we get the following characterizations:

$$a_h(\bar{z}_h, \bar{z}_h) = \langle A_h\bar{Z}, \bar{Z}\rangle , \tag{3.16}$$

where A_h is the stiffness matrix (symmetric) defined by

$$A_h = \{a_{jm}\}_{q\times q} \triangleq \{a_h(w^h_j, w^h_m)\}_{q\times q} ;$$

$$B_h(z_h) = \langle \underline{\beta}(\bar{Z}), \underline{d}\rangle , \quad \{\underline{\beta}(\bar{Z})\}^T = \{\beta(\bar{z}_1),\dots,\beta(\bar{z}_q)\} , \tag{3.17}$$

$$\underline{d} = \{d_j\} = \{ \int_\Omega w^h_1(x)\, dx ,\dots, \int_\Omega w^h_q(x)\, dx \} ;$$

$$\Psi_{\varepsilon h}(\bar{z}_h) = L <\underline{\psi}_\varepsilon(\bar{Z}),\underline{d}> , \tag{3.18}$$

$$\{\underline{\psi}_\varepsilon(\bar{Z})\}^T = \{\psi_\varepsilon(\bar{z}_1),\ldots,\psi_\varepsilon(\bar{z}_q)\} .$$

By (3.16)-(3.18) and in view of the representation

$$\delta\bar{Y}^i_{\varepsilon h}(x) = \sum_{j\in I_q} \bar{u}^i_j\, w^h_j(x) ,$$

where $\bar{u}^i_j \in R$ and $\bar{u}^i_j = \delta G^i_j$ for $j \in I_q \setminus I_p$, the problem (3.14) may be given the form:

Problem $(NP)_\varepsilon$ $(\varepsilon > 0)$: Determine

$$\{\bar{u}^i_1,\ldots,\bar{u}^i_q\} = \arg\inf_{\bar{Z}\in R^q} \{J_\varepsilon(\bar{Z}) \equiv J^i_{\varepsilon h}(\sum_{j\in I_q} \bar{z}_j\, w^h_j)\} \tag{3.19}$$

subject to the constraints $\bar{z}_j = \delta G^i_j$ for $j \in I_q \setminus I_p$,

$$\text{where} \quad J_\varepsilon(\bar{Z}) \triangleq \frac{1}{2}\alpha k <A_h\bar{Z},\bar{Z}> + <\underline{d},\underline{\beta}(\bar{Z}) + L\,\underline{\psi}_\varepsilon(\bar{Z})> + <\underline{C}^i_{\varepsilon h},\bar{Z}> , \tag{3.20}$$

$$\underline{C}^i_{\varepsilon h} = \{c^i_{\varepsilon j}\} ,$$

$$c^i_{\varepsilon j} = a_h(\bar{Y}^i_{\varepsilon h},w^h_j) - ([X_\varepsilon(\delta\bar{Y}_{\varepsilon h})]^i + f^{i+\bar{\alpha}}_\varepsilon , w^h_j)_h - (G^{i+\bar{\alpha}}_2,w^h_j)_{\Gamma'',h} .$$

3.4. Stability and convergence of the approximation schemes

Now we give results on the stability and convergence of the presented approximation method.

Theorem 3.1 [28] . Let the assumptions (A1)-(A9) be satisfied and moreover

(A10) meas $\{ x \in \Omega \mid 0 < \Theta_o(x) < \varepsilon \text{ or } 0 < I_h\Theta_o(x) < \varepsilon \} \leq C\varepsilon$,

where C is a constant independent of ε , h .

Assume that in the case $\alpha \in [0,\frac{1}{2})$ the stability condition

$$\frac{\bar{\bar{\omega}}}{2\bar{\rho}} k (S(h))^2 \leq 1 - \bar{\delta} \tag{3.21}$$

holds with $\bar{\delta} \in (0,1)$ and with a constant $\bar{\bar{\omega}}$ as in (2.38).

Then there exists a constant C independent of ε, h, k , such that the solution $y_{\varepsilon hk}$ of problem $(VI)_{\varepsilon,hk}$ is bounded as follows:

$$\max_{i=0,\ldots,N} \|y^i_{\varepsilon hk}\|_V + \max_{i=0,\ldots,N-1} \|\delta y^i_{\varepsilon hk}\|_V + (\sum_{i=0}^{N-2} k\|\delta^2 y^i_{\varepsilon hk}\|^2_H)^{1/2} \leq C . \tag{3.22}$$

Theorem 3.2 [28]. Assume that the conditions (A1)-(A10) are satisfied and (3.21) holds in the case $\alpha \in [0,\frac{1}{2})$. Let y and $y_{\varepsilon hk}$ be the solutions of (VI) and $(VI)_{\varepsilon,hk}$, respectively. Then there exists a constant C independent of ε, h, k, such that

$$\max_{i=0,\dots,N} \|y^i - y^i_{\varepsilon hk}\|_V + \Big(\sum_{i=0}^{N-1} k \|(y')^i - \delta y^i_{\varepsilon hk}\|_H^2\Big)^{1/2} \leq C h^{1/2}, \quad (3.23)$$

provided $\varepsilon = \varepsilon_o h$, $k \leq k_o h^2$, where $\varepsilon_o > 0$, $k_o > 0$ are arbitrary constants.

Remark 3.4. In the degenerate case (corresponding to $\rho \geq 0$), the solution $y_{\varepsilon hk}$ of $(VI)_{\varepsilon,hk}$ admits the uniform à priori bounds

$$\max_{i=0,\dots,N} \|y^i_{\varepsilon hk}\|_V + \max_{i=0,\dots,N-1} \|\delta y^i_{\varepsilon hk}\|_V \leq C \quad (3.24)$$

with a constant C dependent only upon the bounds on the data. The estimate (3.23) which characterizes the convergence rate of the discrete scheme has to be replaced in the degenerate case (cf., [28]) by

$$\max_{i=0,\dots,N} \|y^i - y^i_{\varepsilon hk}\|_V \leq C h^{1/2} \quad (3.25)$$

provided $\varepsilon = \varepsilon_o h$, $k \leq k_o h^{5/2}$, where $\varepsilon_o > 0$, $k_o > 0$ are arbitrary constants.

4. Numerical tests

Using the introduced approximation schemes, we have performed a number of computational experiments concerned not only with the parabolic but also with degenerate problems of the Stefan type. To solve the minimization problem (3.14), we have applied the SOR method. A quantitative discussion of the obtained numerical results has been given in [29], here we shall put stress rather on qualitative aspects.

Example 4.1 (parabolic two-phase Stefan problem which admits an explicit solution, cf., [5]).

Consider the domain $\Omega = (0,1)\times(0,1) \subset R^2$, with a uniform triangulation $\mathcal{T}_h$ (q - number of nodes in $\mathcal{T}_h$) schematically depicted in Figure 1. Let

$$\rho(\Theta) = \begin{cases} \rho_1 = c_1/k_1 & \text{in } Q_1 \\ \rho_2 = c_2/k_2 & \text{in } Q_2 , \end{cases}$$

with $c_1 = 3$, $c_2 = 2$, $k_1 = 3$, $k_2 = 4$.

Take $L = 1/4$, $\lambda = \rho_1 e^{-4t} - 1$, $\tilde{\lambda} = (\rho_2 - \rho_1) e^{-4t}$. Then the system (2.9)-(2.11) is satisfied by the function

$$\Theta(x_1,x_2,t) = \frac{1}{4}[(x_1)^2 + (x_2)^2 - e^{-4t}] . \qquad (4.1)$$

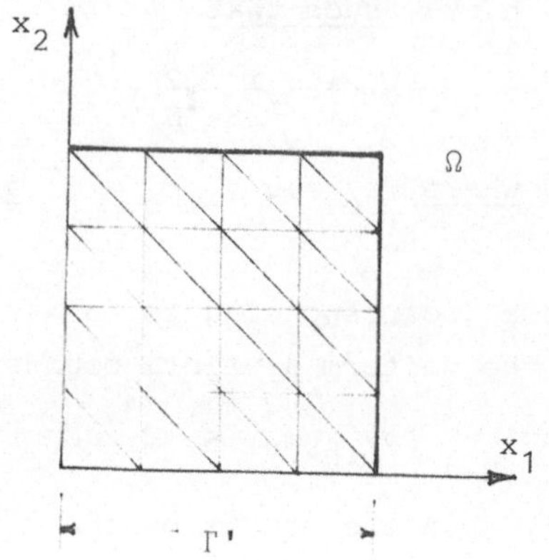

Fig.1. Domain Ω and its triangulation T_h (q=25)

We apply the constructed scheme to solving the following problem which differs in the form of the boundary conditions on Γ (always adjusted so that to be compatible with (4.1)) :

(i) Dirichlet conditions on Γ ;

(ii) Neumann conditions on Γ ;

(iii) mixed type conditions on Γ , with $p = 1$ on Γ' , $p = 0$ on $\Gamma \setminus \Gamma'$.

The initial condition (2.12) is imposed compatibly with (4.1) , as well. The freezing index $\bar{y}$ corresponding to Θ has the form

$$\bar{y}(x_1,x_2,t) = \frac{1}{4}[(x_1)^2+(x_2)^2] t + \frac{1}{16}(e^{-4t}-1) .$$

In all the experiments, we use the uniform discretization of the time interval [0,0.512] with N subintervals.

The terminal distributions of the analytical and computed temperature Θ , referring to the discretization with $q = 81$ and $N = 65$, are depicted in Figures 2a-4a. The relevant terminal locations of the analytical and computed free boundaries are shown in Figures 2b-4b . In all these figures, the analytical solution is represented by the dashed line, against the continuous line which refers to the numerical solution.

In the case (i), where the Dirichlet data are prescribed, there is practically no difference between both solutions, equally as far as the temperature distribution and location of the free boundary are concerned. In (ii) and (iii), despite of a certain difference, the numerical solutions quite accurately reproduce the analytical ones. By solving (iii) at $q = 289$ and $N = 129$, we acquire an easily visible improvement in recovering the temperature and freezing index distribu-

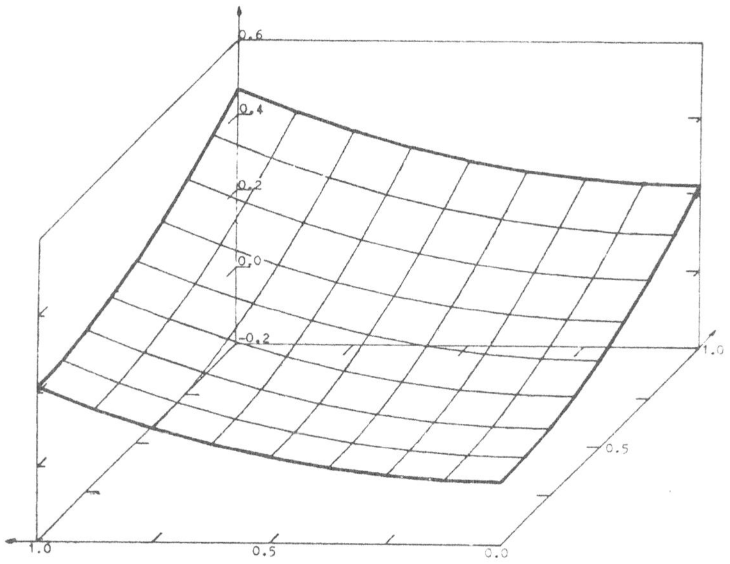

Figure 2a. Example 4.1(i):
Distribution of θ at $t = 0.512$;

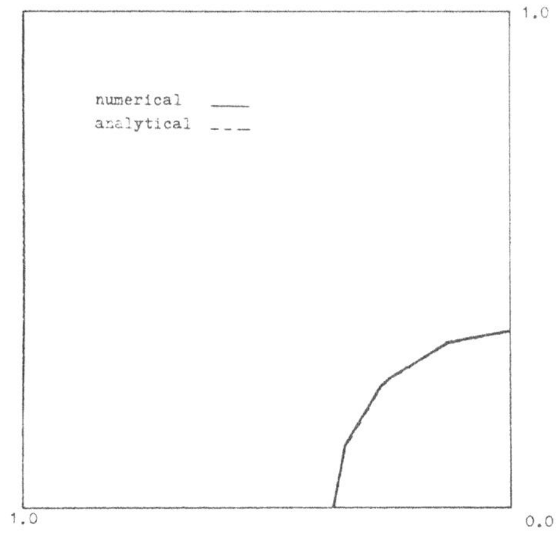

Figure 2b. Example 4.1(i):
Position of the free boundary at $t = 0.512$;

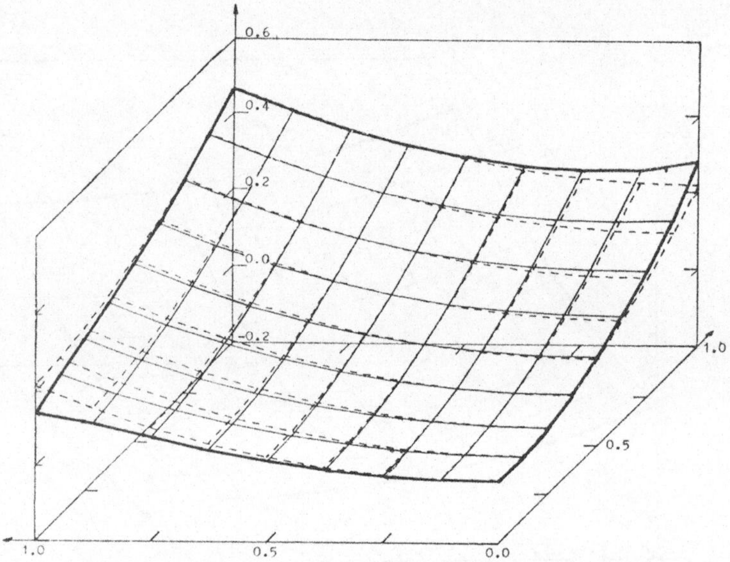

Figure 3a. Example 4.1(ii) : $q = 81$, $N = 65$
Distribution of θ at $t = 0.512$;

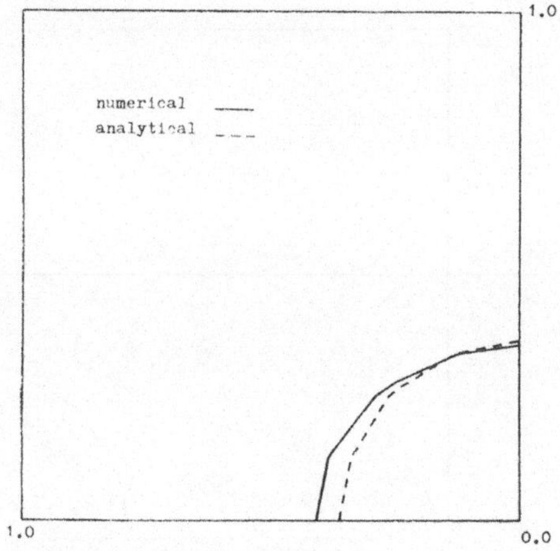

Figure 3b. Example 4.1(ii) :
Free boundary - position at $t = 0.512$;

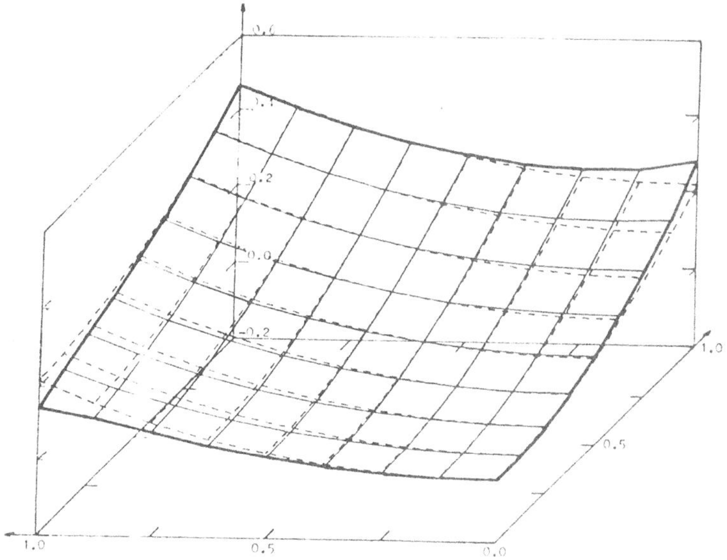

Figure 4a. Example 4.1(iii) : q = 81 , N = 65
Distribution of θ at t = 0.512 ;

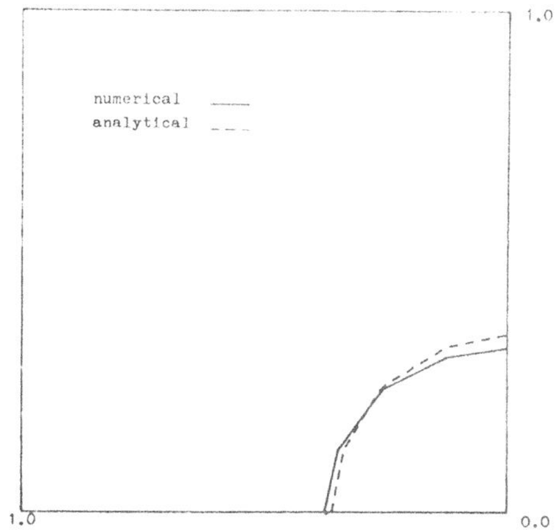

Figure 4b. Example 4.1(iii) : q = 81 , N = 65
Free boundary - position at t = 0.512 ;

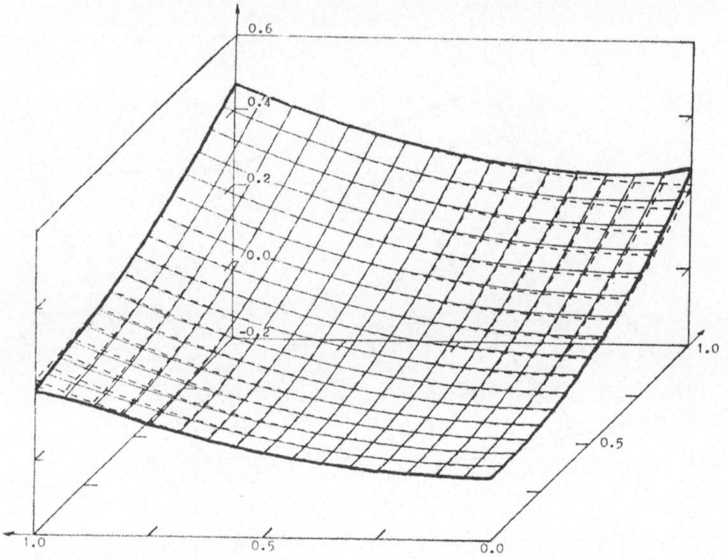

Figure 5a. Example 4.1(iii) : q =289 , N = 129
Distribution of θ at t = 0.512

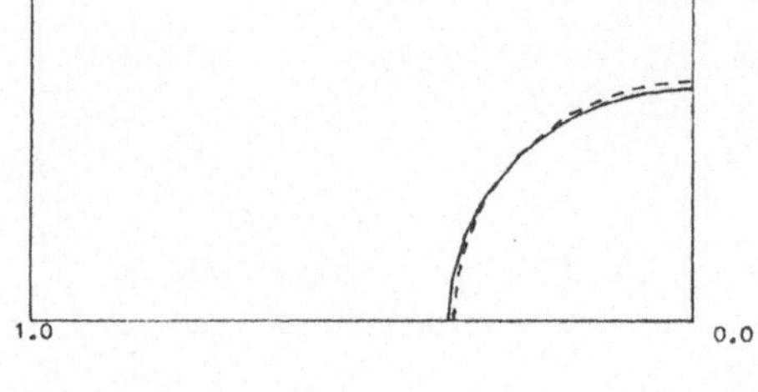

Figure 5b. Example 4.1(iii) : q = 289 , N = 129
Position of the free boundary at t = 0.512

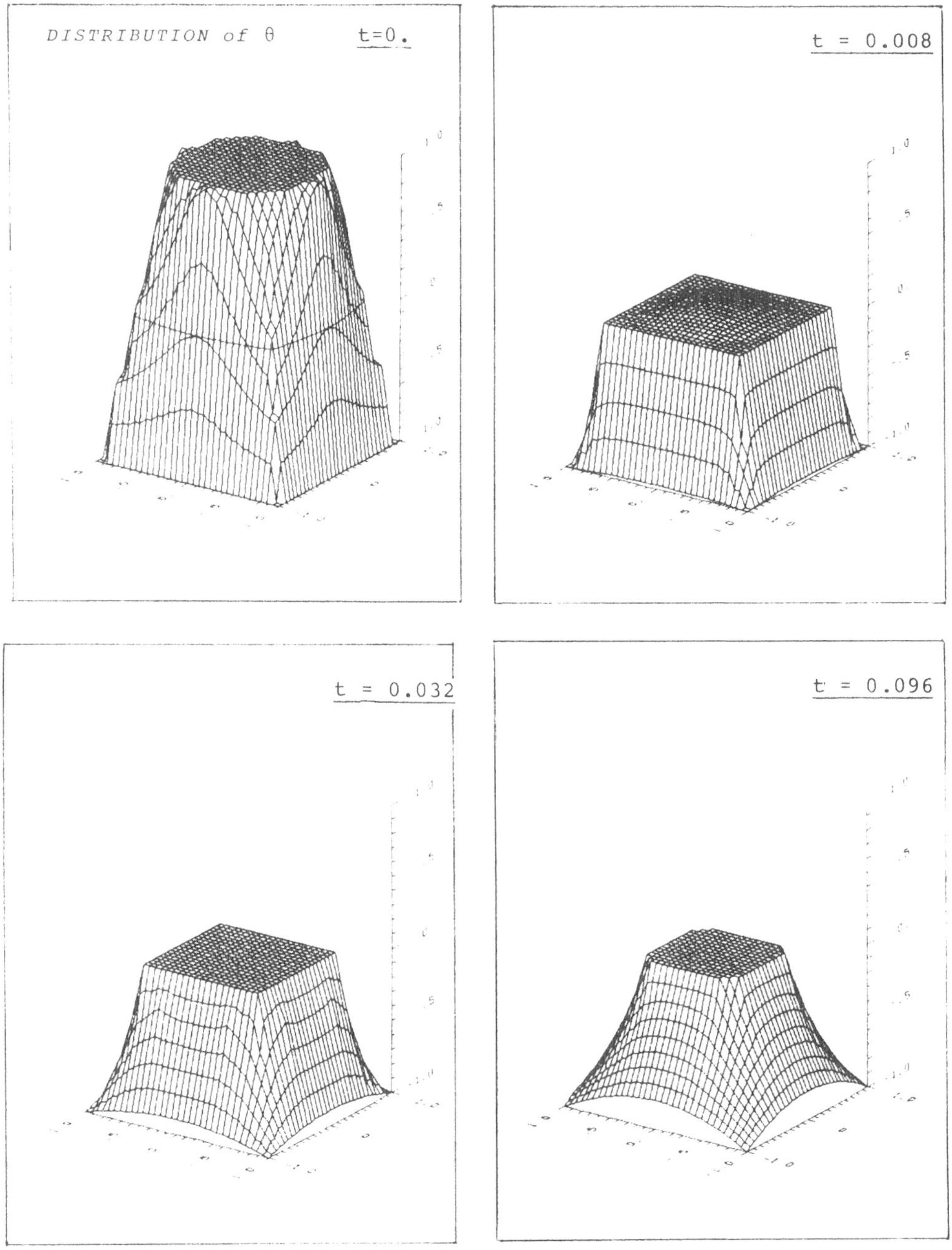

Figure 6. Example 4.2 : Evolution of θ

DIRICHLET BOUNDARY CONDITION ON GAM

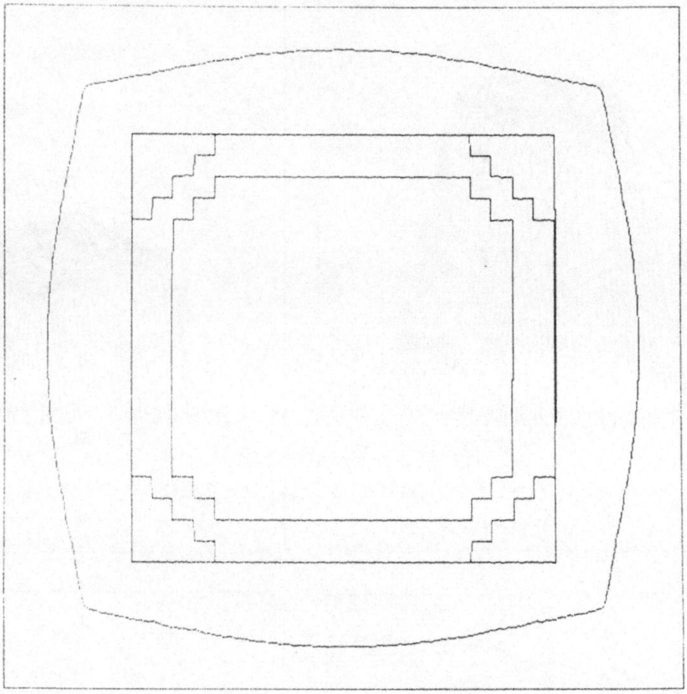

FREE BOUNDARY at t, = 0. (0.032) 0.124

Figure 7. Example 4.2 : Motion of the free boundary

tions as well as in the reconstruction of the free boundary (see Figures 5a,5b). Within our experience, it might be claimed that the "practical" convergence rate of the scheme is of an order higher than theoretically proved $O(h^{1/2})$.

The influence of the regularization parameter ε on the solution is observed to be nonsignificant for $\varepsilon \in [0.001, 0.1]$ as far as the errors measured in the $L^\infty(0,T;L^2(\Omega))$-norm are concerned.

Example 4.2 (a degenerate Stefan problem).

Now we assume $\rho > 0$ in Ω and consider the example 4.1(i) over $\Omega = (-1,1)\times(-1,1)$, with $T = 0.128$. We consider the degenerate problem corresponding to $\tilde{\gamma}(r) \equiv 0$ ($c_1 = c_2 = 0$), with $L = 1$.

Introduce the uniform space and time discretization with the mesh parameters $h = 0.125$ and $k = 0.008$, respectively. We assume that there are no internal sources in Ω and the Dirichlet data are imposed allover Γ so that the free boundary S (level set $\Theta = 0$) do not have common points with Γ.

Together with this problem, we have tested its parabolic regularizations (with $\tilde{\gamma}_\mu(r) = \mu r$, $\mu = 10^{-\alpha}$, $\alpha = 1,2,3$) in order to check the practical influence of the parabolic regularization at $\mu \ll 1$ (theoretically of order $O(\mu^{1/2})$, cf., Remark 3.1). We have not observed any significant changes of the solutions, except for the number of iterations in the SOR process, necessary for achieving an accuracy comparable with that in the strongly parabolic case. In average, we could observe the necessity of carrying out two up to three times more iterations to that purpose (i.e., 15-30 instead of 7-10 pro time step).

Figure 6 shows the time evolution of the distribution of Θ. The corresponding motion of the free boundary S is depicted in Figure 7. The interior phase deformates in time to a square so that its boundary tends to reproduce the shape of Γ (it is of interest, for example, in electrochemical machining processes).

In general, the proposed approach can be recommended as an effective method of determining the solutions also in the case of the degenerate elliptic-parabolic problems of Stefan type.

5. Concluding remarks

The method of numerical solving the two-phase Stefan problems, presented in this paper is applicable within theoretically justified frames under fairly weak hypotheses on the data.

In particular, the requirements imposed on the boundary data are adjusted to the needs of the related problems of boundary control (cf., [27]). A natural extension of the exposed method covers various multi--phase (more than two) situations.

References

[1] D. Blanchard, M. Frémond: The Stefan problem: computing without the free boundary, Int. J. Numer. Meth. Engin., 20 (1984), 757-771.

[2] C.M. Brauner, M. Frémond, B. Nicolaenko: A new homographic approximation to multiphase Stefan problem, in: Free Boundary Problems - Theory and Applications, A. Fasano, M. Primicerio (eds), Research Notes in Math., 79, Pitman, Boston (1983), 365-379.

[3] L. Caffarelli, L.C. Evans: Continuity of the temperature in the two-phase Stefan problem, Arch. Ration. Mech. Anal., 81 (1983), 199-220.

[4] P. Ciarlet: The Finite Element Method for Elliptic Problems, North-Holland, Amsterdam , 1978.

[5] J.F. Ciavaldini: Analyse numérique d'un problème de Stefan à deux phases par une méthode d'elements finis, SIAM J. Numer. Analysis, 12 (1975), 464-487.

[6] L. Čermak, M. Zlamal: Transformation of dependent variables and the finite element solution of nonlinear evolution equations, Int. J. Numer. Meth. Engin., 15 (1980), 31-40.

[7] E. DiBenedetto: Regularity properties of the solutions of an n-dimensional two-phase Stefan problem, Boll. UMI, Suppl., 1 (1980) , 129-152.

[8] G. Duvaut: The solution of a two-phase Stefan problem by a variational inequality, in: Moving Boundary Problems in Heat Flow and Diffusion, J.R. Ockendon, A.R. Hodgkins (eds), Clarendon Press, Oxford (1975), 173-181.

[9] G. Duvaut, J.L. Lions: Les Inéquations en Mecanique et en Physique, Dunod, Paris, 1972.

[10] C.M. Elliott: On the finite element approximation of an elliptic variational inequality arising from an implicit time discretization of the Stefan problem, IMA J. Numer. Anal., 1 (1981), 115-125.

[11] C.M. Elliott, J.R. Ockendon: Weak and Variational Methods for Moving Boundary Problems, Research Notes in Math., 59, Pitman, Boston, 1982.

[12] A. Fasano, M. Primicerio: General free boundary problems for the heat equation, J. Math. Anal. Applications, I: 57 (1977), 694-723; II: 58 (1977), 202-231; III: 59 (1977), 1-14.

[13] M. Frémond: Variational formulation of the Stefan problem - Coupled Stefan problem - Frost propagation in porous media, in: Computational Methods in Non-linear Mechanics, J.T. Oden (ed.), The University of Texas, Austin, 1974.

[14] E.-I. Hanzawa: Classical solution of the Stefan problem, Tohoku Math. J., 33 (1982), 155-172.

[15] U. Hornung: A parabolic-elliptic variational inequality, Manuscripta Mathem., 39 (1982), 155-172.

[16] J.W. Jerome: Approximation of Nonlinear Evolution Systems, Academic Press, New York, 1983.

[17] J.W. Jerome, M.E. Rose: Error estimates for the multidimensional two-phase Stefan problem, Math. of Computation, 39 (160) (1982), 377-414.

[18] N. Kikuchi, Y. Ichikawa: Numerical methods for a two-phase Stefan problem by variational inequalities, Int. J. Numer. Meth. Engin., 14 (1979), 1221-1239.

[19] J.L. Lions: Sur Quelques Questions d'Analyse, de Mécanique et de Contrôle, Presses de l'Université de Montreal, 1976.

[20] A.M. Meirmanov: On the classical solution of the multidimensional Stefan problem for quasilinear parabolic equations, Matem. Sbornik, 112 (1980), 170-192. (in Russian)

[21] G.H. Meyer: Multidimensional Stefan problems, SIAM J. Numer. Analysis, 10 (1973), 522-537.

[22] G.H. Meyer: The numerical solution of multidimensional Stefan problems - a survey, in: Moving Boundary Problems, D.G. Wilson, A.D. Solomon, P.T. Boggs (eds), Academic Press, New York (1978), 73-89.

[23] G.H. Meyer: On the computational solution of elliptic and parabolic free boundary problems, in: Free Boundary Problems, E. Magenes (ed.), Ist. Nazionale di Alta Matematica, Roma (1980), 151-173.

[24] M. Niezgódka, I. Pawlow: Numerical analysis of degenerate Stefan problems, in: Recent Advances in Free Boundary Problems, Control and Cybernetics, 14 (1985), in print.

[25] I. Pawlow: A variational inequality approach to generalized two-phase Stefan problem in several space variables, Annali Matem. Pura Applicata, 131 (1982), 333-373.

[26] I. Pawlow: Approximation of an evolution variational inequality arising from free boundary problems, in: Optimal Control of

Partial Differential Equations, K.-H. Hoffmann, W. Krabs (eds), Birkhäuser-Verlag, Basel (1984), 188-209.

[27] I. Pawlow: Error estimates for Galerkin approximation of boundary control problems for two-phase Stefan type processes, in: Differential Equations and Control Theory, III, INCREST, Bucharest (1983), 129-147.

[28] I. Pawlow: Approximation of variational inequality arising from a class of degenerate multi-phase Stefan problems, submitted to Numerische Mathematik.

[29] I. Pawlow, Y. Shindo, Y. Sakawa: Numerical solution of a multidimensional two-phase Stefan problem, Numerical Functional Anal. Optimization, 7 (1984-85).

[30] L. Rubinstein: The Stefan Problem, AMS, Providence, R.I., 1971.

[31] C. Saguez: Contrôle optimal de systèmes à frontière libre, Thèse, Université Technologique, Compiegne, 1980.

[32] U. Streit: Zur Konvergenzordnung von Differenzenmethoden für die Approximation schwacher Lösungen instationärer Differentialgleichungen, Beiträge zur Numerischen Mathematik, 12 (1984), 181-189.

[33] D.G. Wilson, A.D. Solomon, P.T. Boggs (eds): Moving Boundary Problems, Academic Press, New York, 1978.

Chapter 7

IMPROVEMENT OF AN ALGORITHM FOR THE COMPUTATION OF OPTIMAL CONTROL

Yoshiyuki Sakawa

1. Introduction

In our paper [1], we presented an efficient algorithm for computing the optimal control. We considered a dynamical system defined on a fixed time interval $T=[t_0,t_1]$ and described by

$$dx(t)/dt = f(x(t),\ u(t),t), \quad x(t_0) = x_0 , \tag{1}$$

Here $x(t) \in R^n$ and $u(t) \in R^r$. In the paper [1], we minimized the cost functional

$$J(u) = \theta(x(t_1)) + \int_{t_0}^{t_1} L(x(t),u(t),t)dt \tag{2}$$

by introducing a new cost functional

$$\hat{J}(u) = \int_{t_0}^{t_1} \hat{L}(x(t),u(t),t)dt \tag{3}$$

without the terminal cost, where

$$\hat{L}(x,u,t) = L(x,u,t) + \theta_x(x)f(x,u,t) . \tag{4}$$

Since the above function $\hat{L}(x,u,t)$ is more complicated compared to $L(x,u,t)$, we modify the algorithm given in the paper [1] so that the cost functional (2) can be directly minimized without transformation of the form (3).

2. Modified Algorithm

Let us define the Hamiltonian function

$$H(x,u,\lambda,t) = L(x,u,t) + \lambda f(x,u,t) , \tag{5}$$

where λ is an n-dimensional row-vector. We also define the function

$$K(x,u,\lambda,t;v,C) = H(x,u,\lambda,t) + (u-v)^T C(u-v) , \tag{6}$$

where $C=\mathrm{diag}(c_1,c_2,\ldots,c_r) \geqslant 0$. Let U be a <u>compact</u> and <u>convex</u> subset of R^r, and let Ω be the set of all measurable functions $u: T \to U$ satisfying

$$u(t) \in U \quad \text{for all} \quad t \in T. \tag{7}$$

The modified algorithm is as follows:

<u>Step 0</u>: Select a nominal control $u^0 \in \Omega$. Let $x^0(t)$, $t \in T$ be the corresponding nominal trajectory. Set $i=1$.

<u>Step 1</u>: Computer $\lambda^{i-1}(t)$ by solving the differential equation

$$d\lambda^{i-1}(t)/dt = -H_x(x^{i-1}(t),u^{i-1}(t),\lambda^{i-1}(t),t), \tag{8}$$

$$\lambda^{i-1}(t_1) = \theta_x(x^{i-1}(t_1)). \tag{9}$$

<u>Step 2</u>: Select a nonnegative diagonal matrix C^i properly. Determine $x^i(t)$ and $u^i(t)$, $t \in T$ which satisfy both

$$\begin{aligned} &K(x^i(t),u^i(t),\lambda^{i-1}(t),t;u^{i-1}(t),C^i) \\ &= H(x^i(t),u^i(t),\lambda^{i-1}(t),t) + (u^i(t)-u^{i-1}(t))^T C^i (u^i(t)-u^{i-1}(t)) \\ &= \min_{u \in U} K(x^i(t),u,\lambda^{i-1}(t),t;u^{i-1}(t),C^i) \end{aligned} \tag{10}$$

and the differential equation

$$dx^i(t)/dt = f(x^i(t),u^i(t),t), \quad x^i(t_0) = x_0. \tag{11}$$

<u>Step 3</u>: Calculate

$$J(u^i) = \theta(x^i(t_1)) + \int_{t_0}^{t_1} L(x^i(t),u^i(t),t)dt. \tag{12}$$

If $J(u^i) - J(u^{i-1}) > 0$, make the elements of C^i larger and go to Step 2. Otherwise, set $i: i+1$ and go to Step 1.

Stop the computation if the sequence $\{C^i\}$ is bounded and the sequence $\{u^i(t)\}$ of the constrols converges.

The difference between the above algorithm and the earlier one [1] lies in the terminal condition (9). It is clear that the earlier algorithm in the paper [1] is contained in this algorithm as a special case where $\theta(x(t_1))=0$. It is easily seen that, under the same Assumptions 1, 2, and 3 in the paper [1], all the propositions in [1] still hold true. Since Proposition 3 is the main result of the paper [1], we prove it for the modified algorithm.

3. The Main Result and its Proof

Under the additional assumption that $\theta_x=(\theta_{x_1},\ldots,\theta_{x_n})$ and $\theta_{xx}=(\theta_{x_i x_j})$ are continuous on R^n, we will prove the following proposition (Proposition 3 in [1]).

Proposition: There is a constant $M>0$ independent of i such that the inequality

$$J(u^i)-J(u^{i-1})\leqslant -\frac{1}{2}(4c_i+r-M)\int_{t_0}^{t_1}\|u^i(t)-u^{i-1}(t)\|^2 dt \tag{13}$$

holds for any i, where $r\geqslant 0$ is the minimum eigenvalue of R such that $H_{uu}(x,u,\lambda,t)\geqslant R\geqslant 0$, and c_i is the minimum element of the nonnegative diagonal matrix C^i.

Proof. Define the variables

$$\delta x^i(t)=x^i(t)-x^{i-1}(t),\quad \delta u^i(t)=u^i(t)-u^{i-1}(t).$$

We see that

$$J(u^i)\;-\;J(u^{i-1})=$$

$$=\int_{t_0}^{t_1}[H(x^i,u^i,\lambda^{i-1},t)-H(x^i,u^i-\delta u^i,\lambda^{i-1},t)+$$

$$+H(x^i,u^{i-1},\lambda^{i-1},t)-H(x^{i-1},u^{i-1},\lambda^{i-1},t)-\lambda^{i-1}\delta\dot{x}^i]dt+$$

$$+\theta(x^{i-1}(t_1)+\delta x^i(t_1))-\theta(x^{i-1}(t_1))=$$

$$=\int_{t_0}^{t_1}[H_u(x^i,u^i,\lambda^{i-1},t)\delta u^i-\frac{1}{2}(\delta u^i)^T H_{uu}(x^i,u^i,\lambda^{i-1},t)\delta u^i+$$

$$+H_x(x^{i-1},u^{i-1},\lambda^{i-1},t)\delta x^i+$$

$$+\frac{1}{2}(\delta x^i)^T H_{xx}(x^i,u^{i-1},\lambda^{i-1},t)\delta x^i-\lambda^{i-1}\delta\dot{x}^i]dt+$$

$$+\theta_x(x^{i-1}(t_1))\delta x^i(t_1)+\frac{1}{2}\,\delta x^i(t_1)\theta_{xx}(\bar{x}^i(t_1))\delta x^i(t_1), \tag{14}$$

where

$$u^i=u^i-\alpha_1\delta u^i\ \varepsilon\ U,$$

$$x^i=x^{i-1}+\alpha_2\delta x^i\ \varepsilon\ X,$$

$$\bar{x}^i(t_1)=x^{i-1}(t_1)+\alpha_3\delta x^i(t_1)\ \varepsilon\ X.$$

Here α_i (i=1,2,3) satisfy $0<\alpha_i<1$, and X is defined in the paper [1]. From (6) and (10) we obtain

$$\begin{aligned} & H_u(x^i,u^i,\lambda^{i-1},t)\delta u^i \\ & = K_u(x^i,u^i,\lambda^{i-1},t;u^{i-1},C^i)\delta u^i - 2(\delta u^i)^T C^i \delta u^i \\ & < -2(\delta u^i)^T C^i \delta u^i. \end{aligned} \tag{15}$$

In view of (8) and (9), we obtain

$$\int_{t_0}^{t_1} [H_x(x^{i-1},u^{i-1},\lambda^{i-1},t)\delta x^i - \lambda^{i-1}\delta\dot{x}^i]dt + \theta_x(x^{i-1}(t_1))\delta x^i(t_1) = 0 \tag{16}$$

Since there are constants M_4 and M_4' independent of t and i such that

$$\|H_{xx}(\hat{x}^i(t),u^{i-1}(t),\lambda^{i-1}(t),t)\| \leq M_4,\ \forall t\in T,\ \forall i,$$

$$\|\theta_{xx}(\bar{x}^i(t_1))\| \leq M_4'\ ,\ \forall i\ ,$$

using (15) and (16), we see that

$$\begin{aligned} & J(u^i) - J(u^{i-1}) \\ & \leq \int_{t_0}^{t_1} | -\tfrac{1}{2}(4c_i + r)\|\delta u^i(t)\|^2 + \tfrac{1}{2}M_4\|\delta x^i(t)\|^2 | dt \\ & + \tfrac{1}{2}M_4'\|\delta x^i(t_1)\|^2 . \end{aligned} \tag{17}$$

As in the paper [1], from (11) we obtain

$$\|\delta x^i(t)\|^2 \leq M_5^2(t-t_0)\int_{t_0}^{t_1}\|\delta u^i(\tau)\|^2 d\tau, \tag{18}$$

where M_5 is a constant. Substituting (18) into (17) gives (13), where

$$M = M_5^2(t_1-t_0)[M_4\tfrac{1}{2}(t_1-t_0)+M_4'] . \qquad \text{Q. E. D.}$$

4. Concluding Remarks

An algorithm for the solution of optimal control problems with more general cost functional is presented in this paper. It has been proved that all the results in the paper [1] hold true for this modified algorithm. Namely, the values of the cost functional decrease monotonically at each iteration, and if the sequence $\{C^i\}$ is bounded and the sequence $\{u^i(t)\}$ converges to $u(t)$ almost everywhere on T, then the control $u(t)$ and the corresponding solution $x(t)$ of (1) satisfy the necessary conditions for optimality.

References

[1] Y. Sakawa and Y. Shindo, On global convergence of an algorithm for optimal control, IEEE Trans. on Automatic Control, AC-25 (1980) 1149-1153.

Chapter 8

QUASI-OPTIMAL FEEDBACK FOR LINEAR DIFFERENTIAL GAMES

Naofumi Iwata, Koichi Mizukami, Constantin Vârsan

1. Introduction

In order to find a minimal time feedback law for a linear deterministic control system or to describe optimal feedback laws for terminal linear control stochastic system we are facing difficulties caused by the nonlinear partial differential equations involved. It was shown in [1] that in the matric space consisting of all linear deterministic systems with piecewise analytical coefficients there exists a dense subset of systems for which we may find the minimal time-feedback in an explicit form without appealing to the solution of the partial differential equation. A similar result is true for stochastic linear control systems with terminal functional to be minimized.

In this paper we extend the results of [1] for linear differential games which can be stated as perturbed linear control systems of the form

$$\frac{dx}{dt} = A(t)x + \sum_{i=1}^{m} u_i B_i(t) + f(t), \quad x \in R^n, \quad u \in U \subseteq R^m, \quad t \in [t_o, t_1]$$

$$x(t_o) = x_o \in K \qquad (K \subseteq R^n, \text{ compact}) \tag{s}$$

where $f(.)$ belongs to a fixed set F of bounded measurable functions describing the possibilities of the second player in a geme.

In the stochastic case we consider differential games driven by the following Ito equation

$$dx = [A(t) + \sum_{i=1}^{m} u_i B_i(t) - \sum_{i=1}^{\ell} v_i C_i(t)]dt + D(t)dw(t), t \in [t_o, t_1], \; x \in R^n \tag{S}$$

where $w(.)$ is an n-dimensional Wiener process on the probability space $\{\Omega, \mathfrak{F}, P\}$. In both cases the players u and v are taking values in fixed bounded sets and we are looking for u in feedback form $u(t,x)$ which minimizes a given functional for each strategy $v(.)$ used by the second player. The paper is divided into two parts. The first part concerns the deterministic case and the results stated in Theorems 3 and 4 are completed by an application which is a numerical example from pursuit - evasion games as it appears in [4]. The second part concerns the stochastic case and it has the particular feature that the strategy used by v, which is a nonanticipating process, makes impossible the direct use of the corresponding partial differential equation even as a guide.

In this case, following the idea that the strong and weak solutions of (S) have to generate the same probability measure on $C([t_o,t_1];R^n)$ if $u(t,x)$ is Lipschitz continuous in x, the influence of $v(.)$ in (S) is measured by changing the original measure P into a new one P^v with respect to which the system (S) and the functional to be minimized take the forms derived in [1] for each $v(.)$. It is shown that the analytical form of the quasi-optimal feedback u is independent of the strategy used by the second player and it is the functional which changes its optimal values according to the strategy used by $v(.)$.

In both cases, deterministic and stochastic, the analysis is made from the point of view of the first player u. In the deterministic case the assumptions are rather strong as it can be seen from the numerical example.

2. Deterministic case

2.1. Statement of the problem

Denote by M the set of systems (s) where the matrices $A(t)$, $B(t)=(B_1(t)\dots B_m(t))$ are piecewise analytical on $[t_o,t_1]$, right-hand side continuous at each $t\in[t_o,t_1]$ and such that the controllability matrix

$$\int_{t_o}^{t_1} X(t)B(t)B^*(t)X^*(t)dt \quad \text{is positive definite} \quad (>0) \tag{1}$$

where the nonsinugular matrix $X(t)$, $t\in[t_o,t_1]$ is defined by

$$X(t_o) = I, \ \frac{dX}{dt} = -XA(t), \tag{2}$$

I-being the identity matrix and "*" stands for transposition. The set M is endowed with the metric

$$d(s_1,s_2) = \sup_{t\in[t_o,t_1]} \{ \|A^1(t)-A^2(t)\| + \sum_{i=1}^{m} \| \int_{t_o}^{t} [X^1(s)B_i^1(s) - -X^2(s)B_i^2(s)]ds \| + \| \int_{t_o}^{t} [X^1(s)f^1(s)-X^2(s)f^2(s)]ds \| \} \tag{3}$$

where $\|\cdot\|$ stands for the norm of a matrix, and $(A^i(.),B^i(.),f^i(.))$ defines s_i.

The equality $s_1=s_2$ means $A^1(t)=A^2(t)$, $B^1(t)=B^2(t)$ for all

$t \in [t_o, t_1]$ and $f^1(t) = f^2(t)$ a.e. on $[t_o, t_1]$, which is equivalent to $d(s_1, s_2) = 0$.

The control range set is given by

$$U(t) = \{u \in R^m : \bar{C}_i(t) \leqslant u_i \leqslant \tilde{C}_i(t),\ i=1,\dots,m\},\ t \in [t_o, t_1],$$

where $\bar{C}_i$, $\tilde{C}_i$ may depend on $s \in M$ and fulfil

$$\sup_{t \in [t_o, t_1]} \bar{C}_i(t) < 0, \quad \inf_{t \in [t_o, t_1]} \tilde{C}_i(t) > 0. \tag{4}$$

Denote by $\mathcal{U}$ the set consisting of all piecewise continuous $u: [t_o, t_1] \to R^m$, $u(t) \in U(t)$, $t \in [t_o, t_1]$.

Definition 1

A Borel measurable function $u : [t_o, t_1] \times R^n \to R^m$, $u(t,x) \in U(t)$, is called an admissible feedback control for $s \in M$ if for every $x_o \in R^n$ the corresponding system

$$\frac{dx}{dt} = A(t)x + \sum_{i=1}^{m} u_i(t,x) B_i(t) + f(t),\ x(t_o) = x_o,\ t \in [t_o, t_1]$$

has a unique Filippov solution (see [2]).

Definition 2

An admissible feedback $u(t,x)$ is called optimal for $s \in M$ and compact $K \subseteq R^n$ if for every $x_o \in K$, there exists $T_o \in [t_o, t_1]$ such that the corresponding solution $x(.)$ of (s) with $x(t_o) = x_o$ fulfils $x(T_o) = 0$ and T_o is the minimal time to steer x_o to the origin with resepct to all $u(.) \in \mathcal{U}$.

Definition 3

A system $s \in M$ is solvable if there exists a compact neighourhood K_s of the origin in R^n such that for all $x_o \in K_s$ we can construct time optimal feedback control for s using $X(t)$, $B(t)$ and $f(t)$ only.

In [1] the following results were proved in the case $F = \{0\}$:

Theorem 1

There exists a dense subset $D \subseteq M$ such that every $s \in D$ is solvable.

The set $D \subseteq M$ is constructed as follows. Let $s \in M$ be defined by $A(t)$, $B(t)$, $t \in [t_o, t_1]$, and associate $X(t)$ in (2). Denote $b_i(t) = X(t) B_i(t)$ and since $b_i(t)$ is piecewise analytical function then there exist constant vectors $b_1^i,\dots,b_{n_i}^i$ and piecewise analytical functions $p_j^i(t)$ such that $b_i(t) = \sum_{j=1}^{n_i} p_j^i(t) b_j^i$.

Let $\varepsilon > 0$ be arbitrary but fixed. We divide $[t_o,t_1]$ into N subintervals $I_1,\dots,I_N$ such that each $p_i^j(.)$ maintains the same sign on I_r and $2(\text{meas } I_r)\max\limits_{t,i} s^i(t) \leq \varepsilon$, where $s^i(t) = \sum_{j=1}^{n_i} |p_j^i(t)|$.

Each I_r is divided into n_i subintervals $E_r^{i,j}$, $j=1,\dots,n_i$ such that

$$\int_{I_r} |p_j^i(t)|dt = \int_{E_r^{i,j}} s^i(t)dt,$$

We approximate $b_i(.)$ by $b_i^\varepsilon(.)$, as follows

$$b_i^\varepsilon(t) = s_j^i(t) b_j^i\,, \quad t \in E_r^{i,j},\ j=1,\dots,n_i,\ r=1,\dots,N, \tag{5}$$

$$\text{where} \quad s_j^i(t) = \begin{cases} s^i(t) \text{ sgn } p_j^i(t), & t \in E_r^{i,j} \\ 0 & t \in I_r \setminus E_r^{i,j}. \end{cases}$$

Define $B_i^\varepsilon(t) = X^{-1}(t)\ b_i^\varepsilon(t)$ corresponding to $b_i^\varepsilon(t)$ and introduce $s_\varepsilon \in M$ determined by $A(t)$, $B^\varepsilon(t) = (B_1^\varepsilon(t),\dots,B_m^\varepsilon(t))$. By construction $d(s,s_\varepsilon) \leq \varepsilon$ and the optimal feedback for s_ε is given by

$$u_i^\varepsilon(t,x) = \frac{1}{2}[\tilde{C}_i(t) + \bar{C}_i(t) - (\tilde{C}_i(t) - \bar{C}_i(t)) \text{ sgn} < b_i^\varepsilon(t), X(t)x - h(t) >] \tag{6}$$

where $h(t) = \sum_{i=1}^{m} \int_{t_o}^{t} [\tilde{C}_i(s) + \bar{C}_i(s)] b_i^\varepsilon(s) ds$.

The feedback $u^\varepsilon(t,x)$ is optimal for s_ε and the corresponding compact K_ε, and it is called quasioptimal for s since the following is true (see [1]):

<u>Theorem 2</u>

<u>Let</u> $s \in M$, $x_o \in R^n$. <u>Denote by</u> T_o <u>the minimal time to steer</u> x_o <u>to the origin in</u> R^n <u>using</u> s <u>and</u> $\mathcal{U}$. <u>For each</u> $\varepsilon > 0$ <u>let</u> $u^\varepsilon(t,x)$ <u>be the optimal feedback for</u> (s_ε) <u>given by</u> (6). <u>Then</u> $\lim\limits_{\varepsilon \to 0} x_\varepsilon(T_o) = 0$, <u>where</u> $x_\varepsilon(.)$ <u>is the solution of</u> (s_ε) <u>corresponding to</u> $u_\varepsilon(t,x)$ <u>and to the initial condition</u> x_o.

2.2. <u>Quasi-optimal feedback for differential games</u>

A linear pursuit-evasion differential game we are interested in can be stated as a linear control system with a perturbation

$$\frac{dx}{dt} = A(t)x + \sum_{i=1}^{m} u_i B_i(t) - \sum_{j=1}^{\ell} v_j C_j(t), x(t_o) = x_o, \ t \in [t_o, t_1], x \in R^n, \quad (7)$$

where $u=(u_1,\dots,u_m)$ is the control variable for pursuer taking values in a cube $U_k=\{u \in R^m : |u_i| \leq k, \ i=1,\dots,m\}$ and the control variable for the evader $v=(v_1,\dots,v_\ell)$ takes values in a compact set $V \subseteq R^\ell$.

Denote by $\mathcal{V}$ the set consiting of all measureble $v : [t_o,t_1] \to V$. For each $v(.) \in \mathcal{V}$ the system (7) takes on the form (1) considered in Introduction with $f(t)=-\sum_{j=1}^{\ell} v_j(t)C_j(t)$. We are looking for $u(t,x)$ depending on $f(.)$ such that the system (7) is driven to the orgin in minimal time with respect to $u(.) \in \mathcal{U}$.

We assume that the metrices $A(t)$, $B(t)=(B_1(t),\dots,B_m(t))$ fulfil the conditions required in the definition of M and that $C_j(t)$, $j=1,\dots,\ell$ are piecewise continuous and there exists $\eta \in (0,k)$ such that

$$\{ \sum_{j=1}^{\ell} v_j C_j(t) : v \in V\} \subseteq \{ \sum_{i=1}^{m} u_i B_i(t) : u \in U_k, |u_i| \leq k-\eta\}. \quad (8)$$

Under the condition (8) for any $v(.) \in \mathcal{V}$ there exists $\ell(t)$, $t \in [t_o, t_1]$, measurable such that $|\ell_i(t)| \leq k-\eta$ and

$$\sum_{i=1}^{m} \ell_i(t)B_i(t) = \sum_{j=1}^{\ell} v_j(t)C_j(t). \quad (9)$$

Under the assumption (8) any differential game (7) can be written as

$$\frac{dx}{dt} = A(t)x + \sum_{i=1}^{m} u_i B_i(t) + f(t) \quad (10)$$

where

$$f(t) = - \sum_{i=1}^{\ell} \ell_i(t)B_i(t) \quad (11)$$

where $\ell(t) \in U_k$, $|\ell_i(t)| \leq k-\eta$, $\ell(.)$ measurable and $A(t)$, $B(t)$, $f(t)$ fulfil the conditions in the definition of the set M.

Denote by M_1 the set of systems s in M satisfying (11). It is obvious that any differential game in (7) which satisfies (8) is in M_1.

Similarly to Theorem 1 and Theorem 2 we have the following:

<u>Proposition 1</u>

<u>There exists a dense subset</u> $D \subseteq M_1$ <u>such that every</u> $s \in D$ <u>is solvable</u> (<u>see</u> [1]).

<u>Proof</u>

Let $s \in M_1$ and $\varepsilon > 0$ be arbitrary but fixed, and let $A(t)$, $B(t)$, $f(t)$

define $s \in M_1$.

By hypothesis, $f(t) = -\sum_{i=1}^{m} \ell_i(t)B_i(t)$, where $\ell(t)$ is given as in (11).

Approximate $\ell(.)$ by a piecewise constant $\ell^{\varepsilon}(.)$ such that

$$\ell^{\varepsilon}(t) \in U_k, \quad |\ell_i^{\varepsilon}(t)| \leqslant k-\eta, \tag{12}$$

$$\sum_{i=1}^{m} \int_{t_o}^{t_1} |\ell_i(t)-\ell_i^{\varepsilon}(t)|\,|b_i(t)|dt < \varepsilon \tag{13}$$

where $b_i(t)=X(t)B_i(t)$.

We repeat the construction of $b_i^{\varepsilon}(t)$ using $b_i(t)$ given in Section 2.1.

The construction remains unchanged except that this time we require additionally that on each subinterval I_r the function $\ell^{\varepsilon}(t)$ is constant. Define $f^{\varepsilon}(t) = \sum_{i=1}^{m} \ell_i^{\varepsilon}(t)B_i^{\varepsilon}(t)$, where $B_i^{\varepsilon}(t)=X^{-1}(t)b_i^{\varepsilon}(t)$. Using (13) we obtain

$$\sup_{t\in[t_o,t_1]} \left|\int_{t_o}^{t} X(s)\left[f^{\varepsilon}(s)-f(s)\right] ds\right| \leqslant \sum_{i=1}^{m} \int_{t_o}^{t} |\ell_i^{\varepsilon}(t)-\ell_i(t)|\,|b_i(t)|dt \; +$$

$$\sup_{t\in[t_o,t_1]} \left|\int_{t_o}^{t} \ell_i^{\varepsilon}(s)\left[b_i(s)-b_i^{\varepsilon}(s)\right]ds\right| \leqslant \varepsilon + k \sup_{t\in[t_o,t_1]} \sum_{i=1}^{m} \left|\int_{t_o}^{t} [b_i(s) -\right.$$

$$\left. - b_i^{\varepsilon}(s)]ds\right| \leqslant (1+k)\varepsilon . \tag{14}$$

Define s_{ε} by $A(t)$, $B^{\varepsilon}(t) = (B_1^{\varepsilon}(t),\ldots,B_m^{\varepsilon}(t)$ and $f^{\varepsilon}(t)$.

Using the construction of $b_i^{\varepsilon}(t)$ and (14) yields $d(s,s_{\varepsilon}) \leqslant (2+k)\varepsilon$.

In order to prove that $s_{\varepsilon} \in M_1$ we have to show that the controllability matrix $\int_{t_o}^{t_1} X(t)B^{\varepsilon}(t)(B^{\varepsilon}(t))^* X^*(t)dt$ of s is positive definite and it is done in [1] (see proof of Theorem 1). The system s_{ε} can be written as a linear control system

$$\frac{dx}{dt} = A(t)x + \sum_{i=1}^{m} u_i B_i^{\varepsilon}(t) \tag{15}$$

with the control range set defined by

$$U(t) = \{u \in R^m : -k+\ell_i^{\varepsilon}(t) \leqslant u_i \leqslant \ell_i^{\varepsilon}(t)+k,\ i=1,\ldots,m\}$$

Using Theorem 1 in [1], we find that the system (15) is solvable and the optimal feedback for (15) and for the corresponding compact set K_{ε} is given by

$$u_i^{\varepsilon}(t,x) = \ell_i^{\varepsilon}(t) - k \operatorname{sgn} \langle b_i^{\varepsilon}(t), X(t)x - h(t)\rangle, \quad i=1,\dots,m, \tag{16}$$

where $h(t) = \sum_{i=1}^{m} \int_{t_o}^{t} \ell_i^{\varepsilon}(s) b_i^{\varepsilon}(s)\,ds$. The proof is completed. □

Let $s_\varepsilon \in M_1$ for each $\varepsilon > 0$ with $d(s,s_\varepsilon) < \varepsilon$ such that the feedback $u^{\varepsilon}(t,x)$ in (16) is optimal for s_ε and for the corresponding compact neighbourhood K_ε. For s given by (7) the optimal feedback $u^{\varepsilon}(t,x)$ defined by (16) for s_ε fulfils the following:

<u>Proposition 2</u>

<u>Let</u> $s \in M$ <u>be defined by the differential game</u> (7), <u>where</u> (8) <u>is filled</u>. <u>Denote by</u> T_o <u>the minimal time to steer</u> x_o <u>to the origin in</u> R^n <u>with respect to</u> $\mathcal{U}$ <u>using s, with a fixed</u> $v(.) \in \mathcal{V}$. <u>For each</u> $\varepsilon > 0$ <u>let</u> s_ε <u>be defined by</u> (15) <u>and let</u> $u^{\varepsilon}(t,x)$ <u>be the corresponding optimal feedback given by</u> (16). <u>Then</u> $\lim_{\varepsilon \to 0} x_\varepsilon(T_o) = 0$, <u>where</u> $x_\varepsilon(.)$ <u>is the solution of</u> (15) <u>corresponding to</u> $u_\varepsilon(t,x)$ <u>and</u> $x(0)=x_o$.

The proof of Proposition 2 is the same as that of Theorem 2 in [1] since the differential game (7) is converted into a linear control system in [1] using the condition (8).

The feedback $u^{\varepsilon}(t,x)$ defined by (16) and used in the above propositions has an unpleasant feature. Namely it is not enough to know the strategy used by evader via (9) but, in addition, we have to approximate this strategy by another $\ell^{\varepsilon}(.)$ which satisfies (12) and (13).

It is necessary as long as the subset D in Proposition 1 is dense in M_1 in the sense of the metric d defined by (3).

In the sequel we shall restate the density property in Proposition 1 using the simpler metric d given in [1] for control systems and as a consequence the feedback $u^{\varepsilon}(t,x)$ will not be dependent on $\ell^{\varepsilon}(.)$ but on $\ell(.)$.

Any differential game (7) fulfiling (8) can be written as $\frac{dx}{dt} = A(t)x + \sum_{i=1}^{m} u_i B_i(t)$ with the control range set $U(t)$ for u depending on the strategy used by v via (11).

$$U(t) = \{u \in R^m,\ -k+\ell_i(t) \leqslant u_i \leqslant \ell_i(t)+k,\ i=1,\dots,m\}.$$

In this way, the differential game (7) satisfying (8) can be considered as defining a subset $\tilde{M}_1$ of control systems of the form studied in [1] for which we do not fix the functions $\bar{C}_i(t)$, $\tilde{C}_i(t)$ in the definition of the control set $U(t)$ but preserve the condition (4) and impose an uniform boundedness condition on the functions $\bar{C}_i(.)$, $\tilde{C}_i(.)$. The metric d in (3) can be replaced by

$$\tilde{d}(s_1,s_2) = \sup_{t\in[t_o,t_1]} \{ \|A^1(t)-A^2(t)\| + \sum_{i=1}^{m} \| \int_{t_o}^{t} [X^1(s)B_i^1(s)-X^2(s)B_i^2(s)]ds \| \}$$

for $s_1,s_2 \in \tilde{M}_1$, which is the same as the metric used in [1] for deterministic control systems.

The result in Proposition 1 can be restated as follows:

Theorem 3

Let the differential game (7) fulfils (8). Then for each $\varepsilon > 0$, for each strategy $v(.) \in \mathcal{V}$ used by v and for the corresponding control system $s \in \tilde{M}_1$ there exists $s_\varepsilon \in \tilde{M}_1$ with $\tilde{d}(s,s_\varepsilon) < \varepsilon$ and every s_ε is solvable with the feedback

$$\tilde{u}_i^\varepsilon(t,x) = \ell_i(t) - k \operatorname{sgn} \langle b_i^\varepsilon(t), X(t)x-h(t)\rangle, \quad i=1,\dots,m,$$

where $h(t) = \sum_{i=1}^{m} \int_{t_o}^{t} \ell_i(s)b_i^\varepsilon(s)ds$, $\ell_i(.)$ *fulfils* (9), $b_i^\varepsilon(t)=X(t)B_i^\varepsilon(t)$, *and* $(A(.), B_i^\varepsilon(.))$ *defines* s_ε.

The proof parallels that of Proposition 1 except for the approximate $\ell^\varepsilon(.)$.

The optimal feedback $\tilde{u}(t,x)$ defined in Theorem 3 for s_ε is called quasi-optimal for s and it fulfils in addition the following:

Theorem 4

Let $s \in \tilde{M}_1$ be defined by differential game (7) which fulfils (8). Denote by T_o the minimal time to steer x_o to the origin in R^n with respect to $u(.) \in \mathcal{U}$ and for a fixed $v(.) \in \mathcal{V}$. For $\varepsilon > 0$ let $s_\varepsilon \in \tilde{M}_1$ and $\tilde{u}^\varepsilon$ be the quasi-optimal feedback given in Theorem 3. Then $\lim_{\varepsilon\to 0} x_\varepsilon(T_o)=0$, *where $x_\varepsilon(.)$ is the solution in s_ε corresponding to $\tilde{u}^\varepsilon$ and* $x(0)=x_o$.

2.3. Example

We consider a pursit-evasion game as in [4] described by the following system

$$\frac{dx_p}{dt} = \begin{bmatrix} 0,1 \\ 1,0 \end{bmatrix} x_p + u_1 \begin{bmatrix} 1 \\ 1 \end{bmatrix} + u_2 \begin{bmatrix} 0 \\ 1 \end{bmatrix}, \quad x_p(0) = \begin{bmatrix} 0 \\ 0 \end{bmatrix}, \quad t \in [0,1.4]$$

$$\frac{dx_e}{dt} = \begin{bmatrix} 0,1 \\ 1,0 \end{bmatrix} x_e + v_1 \begin{bmatrix} 1 \\ 0 \end{bmatrix} + v_2 \begin{bmatrix} 0 \\ 1 \end{bmatrix}, \; x_e(0) = \begin{bmatrix} 10 \\ 10 \end{bmatrix}, \; t \in [0,1.4] \qquad (\alpha)$$

where the variable u for pursuer fulfils $|u_i| \leqslant 20$, $i=1,2$, and the strategy for v is $v=\begin{bmatrix} 1 \\ 1 \end{bmatrix}$.

The goal is to obtain u_1, u_2 in feedback form minimizing T such that $x_p(T)-x_e(T)=\begin{bmatrix} 0 \\ 0 \end{bmatrix}$. Denote $x=x_p-x_e$. It follows

$$\frac{dx}{dt} = \begin{bmatrix} 0,1 \\ 1,0 \end{bmatrix} x + u_1 \begin{bmatrix} 1 \\ 1 \end{bmatrix} + u_2 \begin{bmatrix} 0 \\ 1 \end{bmatrix} - \begin{bmatrix} 1 \\ 1 \end{bmatrix}, \; x(0) = -\begin{bmatrix} 10 \\ 10 \end{bmatrix} \qquad (\beta)$$

and the original problem is equivalent to finding u_1, u_2 in a feedback form minimizing T such that $x(T)=\begin{bmatrix} 0 \\ 0 \end{bmatrix}$.

The condition (11) is fulfilled for (β) by taking $\ell_1=1$, $\ell_2=0$ and the system (β) becomes

$$\frac{dx}{dt} = \begin{bmatrix} 0,1 \\ 1,0 \end{bmatrix} x + u_1 \begin{bmatrix} 1 \\ 1 \end{bmatrix} + u_2 \begin{bmatrix} 0 \\ 1 \end{bmatrix}, \; x(0) = -\begin{bmatrix} 10 \\ 10 \end{bmatrix} \qquad (\gamma)$$

with the control range set defined by

$$-19 \leqslant u_1 \leqslant 21 \; , \qquad |u_2| \leqslant 20.$$

The system (γ) is controllable and there exist

$u_1 : [0,1.4] \to [-19,21]$, $u_2 : [0,1.4] \to [-20,20]$ piecewise

continuous and $T \in [0,1.4]$ such that the corresponding solution $x(.)$ of (γ) fulfils $x(T)=\begin{bmatrix} 0 \\ 0 \end{bmatrix}$; it ensures that the class of admissible controls is nonempty. The assumptions of Theorem 3 are fulfilled for (γ) and we take the values as follows:

$$I_r = 0.02, \quad r=1,\dots,70$$

judgement distance =0.04.

The results are:

final time: $t_j=1.29$

final state: $x_e(t_f)=\begin{bmatrix} 20.177 \\ 19.889 \end{bmatrix}$, $x_p(t_f)=\begin{bmatrix} 20.198 \\ 19.880 \end{bmatrix}$

winner: pursuer.

See the following figures.

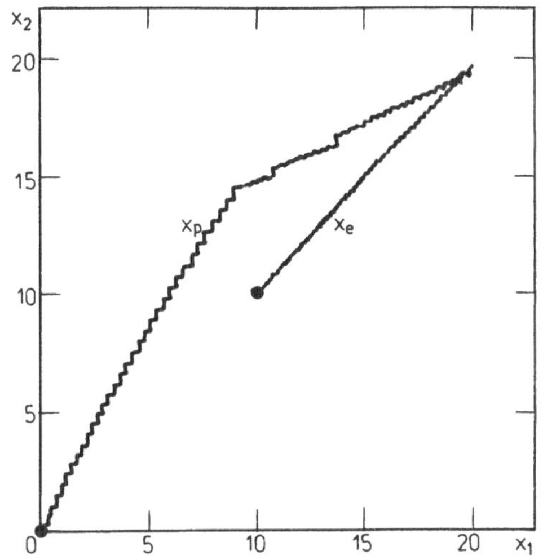

Fig.1 Trajectories of x_p and x_e .

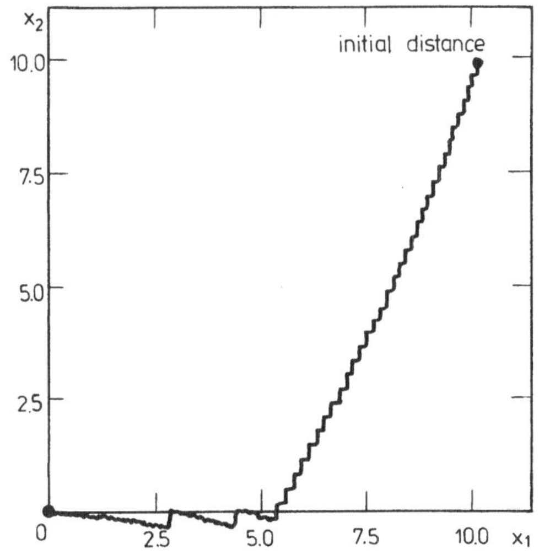

Fig. 2 Distance between x_p and x_e .

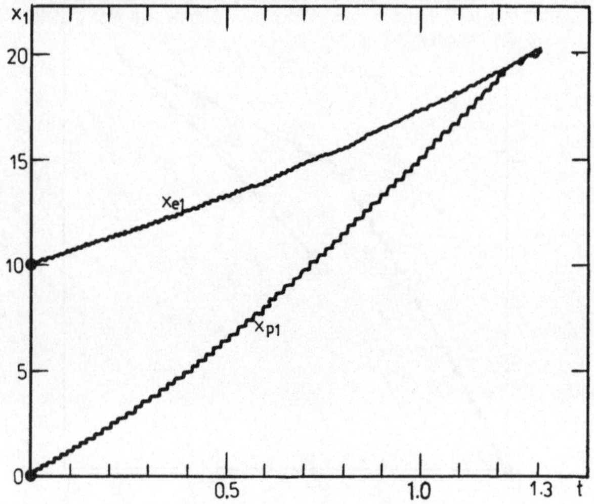

Fig.3 State variations of x_{p1} and x_{e1} againts time t.

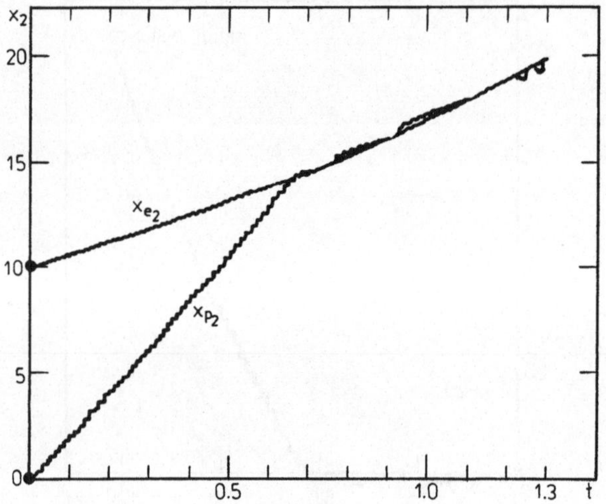

Fig.4 State variations of x_{p_2} and x_{e_2} againts time t.

3. Stochastic case

A density property, similar to that for the deterministic case, will be studied for stochastic linear differential games described by Ito equations

$$dx = \left[A(t)x + \sum_{i=1}^{m} u_i B_i(t) - \sum_{i=1}^{\ell} v_i C_i(t)\right]dt + D(t)dw(t), \quad t \in [t_o, t_1], \quad x \in R^n,$$

where $w(t)$, $t > t_o$, is an n-dimensional standard Wiener process ($w(t_o) = 0$) on the probability space $\{\Omega, \mathcal{F}, P\}$.

The control range set $U(t)$ for control u is defined as in the deterministic case (see(3)), and for the second player v it is a given bounded set $V \subseteq R^{\ell}$.

Using the mapping $(t,y) \to (t, X(t)x)$, $t \in [t_o, t_1]$, $x \in R^n$, $y \in R^n$, where

$$X(t_1) = I, \quad \frac{dX}{dt} = -XA(t), \quad t \in [t_o, t_1],$$

any linear stochastic system is converted into an equivalent one with $y(t_1) = x(t_1)$ for which the dynamics is described by

$$dy = \left[\sum_{i=1}^{m} u_i b_i(t) - C(t)v\right]dt + H(t)dw(t), \quad t \in [t_o, t_1], \quad y(0) = X(t_o)x(0) \qquad \text{(S)}$$

where $b_i(t) = X(t)B_i(t)$, $C(t) = X(t)\left[C_1(t) \ldots C_{\ell}(t)\right]$, $H(t) = X(t)D(t)$,

and the matrix $H(t)$ is nonsingular provided that $D(t)$ is nonsingular. Without any loss of generality we shall study the set of systems of the form (S). Let N be the set consisting of all systems (S) with piecewise analytical vector functions $b_i(t)$, $i=1,\ldots,m$, the matrix function $C(.)$ being bounded and measurable and $H(t)$ being continuous matrix function with det $H(t) \neq 0$, for any $t \in [t_o, t_1]$.

On N we consider the following metric

$$d(S_1, S_2) = \max_{t \in [t_o, t_1]} \Big\{ \sum_{i=1}^{m} \Big\| \int_{t_o}^{t} [b_i^1(s) - b_i^2(s)]ds \Big\| + \|H^1(t) - H^2(t)\| + \Big\| \int_{t_o}^{t} [C^1(s) - C^2(s)]ds \Big\| \Big\}$$

where S_j is defined by $b_i^j(t)$, $i=1,\ldots,m$, $H^j(.)$, $C^j(.)$ and $\|\cdot\|$ is the norm of a $(n \times \ell)$ matrix considered as a vector in $R^{n\ell}$.

The admissible class of controls $\mathcal{U}$ consists of all Borel measurable $u(t,y) : [t_o, t_1] \times R^n \to R^m$ such that $u(t,y) \in U(t)$, $t \in [t_o, t_1]$;

the admissible class of controls $\mathcal{V}$ consists of all $v(t,\omega):[t_o,t_1]\times\Omega\to V$ measurable and nonanticipating with respect to the σ-algebras $\mathcal{F}_t=\sigma\{w(s):t_o\leq s\leq t\}$. Since we are going to work with admissible class $\mathcal{U}$ and $\mathcal{V}$ it is necessary to define the meaning of weak solutions in (S) and according to this the functional to be minimized. For each $v(.)\in\mathcal{V}$, $u(.)\in\mathcal{U}$ define

$$y^v(t)=y-\int_{t_o}^{t} C(s)v(s,\omega)ds+\int_{t_o}^{t} H(s)dw(s),\quad t\in[t_o,t_1]$$

$$p^u(t,y)=\sum_{i=1}^{m} H^{-1}(t)b_i(t)u_i(t,v).$$

It follows that using Girsanov's theorem (see [3]) the process $y(.)$ fulfils the Ito equation

$$dy=\Big[\sum_{i=1}^{m} v_i(t,y)b_i(t)-C(t)v(t,\omega)\Big]dt+H(t)dw^{u,v}(t),$$

$$y(t_o)=y$$

where $w^{u,v}(t)$, $t\in[t_o,t_1]$, is a Wiener process on the probability space $\{\Omega,\mathcal{F},P^{u,v}\}$, with $P^{u,v}=k^{u,v}(t_1)P$, and

$$k^{u,v}(t_1)=\exp\{\int_{t_o}^{t_1} p^u(t,y^v(t))dw(t)-\frac{1}{2}\int_{t_o}^{t_1}|p^u(t,y^v(t))|^2dt\} \tag{17}$$

The process $y^v(t)$, $t\in[t_o,t_1]$, on the probability space $\{\Omega,\mathcal{F},P^{u,v}\}$ will be called the weak solution of (S) corresponding to $u(.)\in\mathcal{U}$ $v(.)\in\mathcal{V}$ and initial condition $y(t_o)=y\in R^n$. When $u(.)\in\mathcal{U}$ is such that $u(t,y)$ is Lipschitz continuous in y uniformly with respect to $t\in[t_o,t_1]$, $(u(.)\in\mathcal{U}_1)$ then for $v(.)\in\mathcal{V}$ the Ito solution $y^{u,v}(.)$ of (S) is called a strong solution and $y^{u,v}(.)$ on $\{\Omega,\mathcal{F},P\}$, $y^v(.)$ on $\{\Omega,\mathcal{F},P^{u,v}\}$ generate the same probability on $C([t_o,t_1];R^n)$ provided that the original probability P is replaced by $P^v=k^v(t_1)P$, where

$$k^v(t_1)=\exp\{\int_{t_o}^{t_1}\langle\tilde{C}(t)v(t,\omega),dw(t)\rangle-\frac{1}{2}\int_{t_o}^{t_1}|\tilde{C}(t)v(t,\omega)|^2dt\},$$

$$\tilde{C}(t)=H^{-1}(t)C(t) \tag{18}$$

In the sequal we shall clarify this assertion.

It can be seen that $y^v(.)$ has an equivalent form

$$y^v(t)=y+\int_{t_o}^{t} H(s)dw^v(s), \tag{19}$$

$$\text{where}\quad w^v(t)=w(t)-\int_{t_o}^{t}\tilde{C}(s)v(s,\omega)ds,\quad t\in[t_o,t_1]$$

is a Wiener process on the probability space $\{\Omega,\mathcal{F},P^v\}$ and the equation (S) becomes

$$dy=\sum_{i=1}^{m}u_i(t,y)b_i(t)dt+H(t)d\tilde{w}^{u,v}(t),\quad y(t_o)=y \tag{S'}$$

or

$$dy=\sum_{i=1}^{m}u_i(t,y)b_i(t)dt+H(t)dw^v(t),\quad y(t_o)=y \tag{S"}$$

according to the weak solution or to the strong one we may define, where $\tilde{w}^{u,v}(.)$ is obtained from $w^{u,v}(.)$ by replacing $w(.)$ by $w^v(.)$. Denote $\tilde{P}^{u,v}=k^{u,v}(t_1)P^v$. Since $\tilde{w}^{u,v}(.)$ is a Wiener process on $\{\Omega,\mathcal{F},\tilde{P}^{u,v}\}$ it follows that when $u(.)\in\mathcal{U}_1$ the strong solutions of (S') and (S") generate the same probability on $C([t_o,t_1];R^n)$ and it clarifies the substitution of the original probability P by P^v in the analysis we have to do with the functionals depending on the solution of (S). In this way the terminal functional in [1] (see (19)) in stochastic case has to be replaced by

$$J(u,v)=E^v|y^{u,v}(t_1)|^2 \tag{20}$$

where E^v is the expection with respect to P^v, and $y^{u,v}(.)$ is the Ito solution of (S") when $u(.)\in\mathcal{U}_1$.

Accordingly, when $u(.)\in\mathcal{U}$ we have to use (S') and the weak solution $y^v(.)$ in (19) and the functional

$$\tilde{J}(u,v)=\tilde{E}^{u,v}|y^v(t_1)|^2 \tag{21}$$

where $\tilde{E}^{u,v}$ is the expectation with respect to the probability $\tilde{P}^{u,v}=k^{u,v}(t_1)k^v(t_1)P$, and $k^{u,v}(t_1)$, $k^v(t_1)$ are defined in (17) and (18).

Definition 4

$\tilde{u}(.)\in\mathcal{U}$ is called optimal if $\tilde{J}(\tilde{u},v)\leq\tilde{J}(u,v)$ for any $u(.)\in\mathcal{U}$ and for $v(.)\in\mathcal{V}$; $\tilde{u}_\varepsilon(.)\in\mathcal{U}$, $\varepsilon\in(0,1)$ is called quasi-optimal for $S\in N$ if it is optimal for some $S_\varepsilon\in N$ where $d(S,S_\varepsilon)<\varepsilon$.

The following results have been proved in [1] for the case $C(t)=0$:

Theorem 5 (see [1])

There exists a dense subset $D\subseteq N$ such that for each $S\in D$ one may define the corresponding optimal feedback law in explicit form, i.e. for any $S\in N$ there exists a quasi-optimal feedback law.

Definition 5

A function $b(t):[t_o,t_1] \to R^n$ is called piecewise constant in direction (p.c.d.) if there exist a partition $t_o=\tau_o<\tau_1<\ldots<\tau_\ell=t_1$ and an orthonormal basis $e_1,\ldots,e_n \in R^n$ such that $b(t)=c_i(t)e_i$ for $t\in[\tau_i,\tau_{i+1})$ where $e_i\in\{e_1,\ldots,e_n\}$ and $c_i(.)$ is a scalar function.

The result in the above theorem is essentially based on the following:

Lemma 1

Let $b_i(t)$, $i=1,\ldots,m$, be p.c.d. with respect to a common partition $t_o=\tau_o<\tau_1<\ldots<\tau_d=t_1$ and to an orthonormal basis $\{e_1,\ldots,e_n\}$ such that $b_i(t)$ is continuous on each $[\tau_j,\tau_{j+1})$. Then $\tilde{u}_i(t,y)=-\mathrm{sgn}\langle b_i(t),y\rangle$ $i=1,\ldots,m$, is an optimal feedback for (S) and functional (21) for $v=0$ w.r.t. all Borel measurable $u(t,y):[t_o,t_1]\times R^n \to \prod_1^m [-1,1]$.

Similar results are true for the differential game (S) and the functional (21).

Theorem 6

There exist a dense subset $D\subseteq N$ such that for each $S\in D$ one may define the corresponding optimal feedback law in an explicit form and independent of the strategy $v(.)\in\mathcal{V}$ (see (26)); i.e. for any $S\in N$ there exists a quasi-optimal feedback law which does not depend on $v(.)\in\mathcal{V}$.

Proof

Let $S\in N$ and $\varepsilon>0$ be arbitrary but fixed. Denote $I_1,\ldots,I_p$ the intervals of analycity of the functions $b_i(t)$, $i=1,\ldots,m$, defining (S). As in the deterministic case (see the proof of Theorem 1) we get that there exist $b_i^\varepsilon(t),\ldots,b_m^\varepsilon(t)$, p.c.d. with respect to a common partition and basis such that $b_i^\varepsilon(.)$ is continuous on each subinterval of this partition and

$$\max_{t\in[t_o,t_1]} \left\| \int_{t_o}^{t} [b_i(s)-b_i^\varepsilon(s)]ds \right\| < \frac{\varepsilon}{m}, \int_{t_o}^{t_1} [b_i(t)-b_i^\varepsilon(t)]dt=0 \tag{22}$$

Define $S_\varepsilon\in N$ by $b_1^\varepsilon(t),\ldots,b_m^\varepsilon(t)$, $C^\varepsilon(t)=C(t)$, $H^\varepsilon(t)=H(t)$, $t\in[t_o,t_1]$ and form (22) we have $d(S,S_\varepsilon)<\varepsilon$.

The system S_ε written in the form S'_ε (see (S')) for any fixed $v(.)\in\mathcal{V}$ fulfils all assumptions of Lemma 1, except that the control range set is $\mathcal{U}(t)$, $t\in[t_o,t_1]$, which can be converted into the cube

$\prod_{i=1}^{m}[-1,1]$ by the following control and state transformations

$$\ell_i = 2(\tilde{C}_i(t)-\bar{C}_i(t))^{-1}[u_i-\tfrac{1}{2}(\tilde{C}_i(t)+\bar{C}_i(t))],\ u\in\mathcal{U}(t),$$
$$z = y-h(t),\ h(t_1)=0,\ \frac{dh}{dt}=\frac{1}{2}\sum_{i=1}^{m}(\tilde{C}_i(t)+\bar{C}_i(t))b_i^{\varepsilon}(t). \tag{23}$$

Using (S'_ε) we obtain

$$d_z = \sum_{i=1}^{m}\ell_i(t,z)\bar{b}_i^{\varepsilon}(t)dt+H(t)d\tilde{w}^{u,v}(t),\ z(t_o)=y-h(t_o),\ t\in[t_o,t_1],$$
$$z(t_1)=y(t_1),\ \ell_i(t,z)\in[-1,1],\ \bar{b}_i^{\varepsilon}(t)=\frac{1}{2}(\tilde{C}_i(t)-\bar{C}_i(t))b_i^{\varepsilon}(t). \tag{24}$$

The system (24) and the functional $\tilde{J}(u,v)=\tilde{E}^{u,v}|z^v(t_1)|^2$ fulfil the assumptions of Lemma 1 for each $v(.)\in\mathcal{V}$ and we get the optimal feedback

$$\tilde{\ell}_i(t,z) = -\text{sgn}<\bar{b}_i^{\varepsilon}(t),z>=-\text{sgn}<b_i^{\varepsilon}(t),z>,\ i=1,\dots,m \tag{25}$$

Now using (25) and (23) we get the optimal feedback for (S'_ε) and the functional (21)

$$\tilde{u}_i(t,y) = \frac{1}{2}[\tilde{C}_i(t)(1+\tilde{\ell}_i(t,y-h(t))) + \bar{C}_i(t)(1-\tilde{\ell}_i(t,y-h(t)))] \tag{26}$$

$i=1,\dots,m$, where $\tilde{\ell}_i$ are defined in (25).
The proof is completed. □

4. Concluding remarks

There was analyzed the possibility to find the optimal feedback for linear differential games without appealing to the solution of the partial differential equation involved.

In both cases, deterministic and stochastic, the analysis is made from the point of view of the first player u. In the deterministic case the quasi-optimal feedback requires the knowledge of the strategy used by the second player as it appears in Theorem 3. It seems that this condition could be weakened by enlarging the class of admissible feedbacks.

In the stochastic case we allow that the second player v(.) uses nonanticipating processes as the admissible strategy and it makes impossible the direct use of the corresponding parabolic differential equation involved in the study of the optimality.

In this case it is shown that the analytical form of the quasi-

-optimal feedback for the first player u is independent of the strategy used by the second player v and it is a functional which changes its optimal values according to the strategy used by v; it would be interesting to see this particular feature on numerical examples.

References

[1] K. Mizukami, C. Vârsan, Quasi-Optimal Feedback Laws, Rev. Roum. Math. Pures Appl., tom XXVII, Nr 10 (1982), pp. 1027-1051.

[2] A.F. Filippov, Right-Hand Side Discontinuous Differential Equations, Mat. Sbornik, 51 (1960), pp. 99-126 (in Russian).

[3] W. Fleming, R. Rishel, Deterministic and Stochastic Optimal Control, Springer-Verlag, Berlin, 1975.

[4] N. Iwata, The Pursuit-Evasion Differential Game and the Quasi-Optimal Theory, thesis, University of Hiroshima, Horishima, 1980.

Chapter 9

SUBOPTIMAL STRATEGIES FOR NASH NONLINEAR DIFFERENTIAL GAMES

Maciej Krawczak

1. Introduction

A considerable effort has been devoted to the theoretical study on differential games but very little to the development of approximation of the solution or computational techniques to solve this kind of problems. It is well-known that optimal control problems can be treated as a particular case of differential games with only one player. In general, numerical solutions of optimal control problems are very difficult and this subject is still under development. It seems that the numerical solution of differential games is a much more difficult. As for as the author is aware, only few works have been reported about the approximation and computation of differential game solutions.

The Nash equilibrium solution of a two-person, nonzero-sum, linear differential game with two quadratic cost functions can be expressed in terms of the solution of coupled Riccati matrix differential equations. In [18], the solution of these equations is assumed to be an analytic function of a small parameter and a truncated series solution is considered. In [11], a singularly perturbed differential Nash game is considered. The Newton-Raphson method in function space and the gradient optimization of the Hamiltonian function with respect to strategies in an almost linear pursuit-evasion game with a quadratic criterion is investigated in [32]. It seems, that these methods can be used to solve nonlinear Nash games. An algorithm to approximate open-loop Nash equilibrium controls for nonlinear games is proposed in [30]. The algorithm transforms the game into a two-level hierarchical game with an upper level referee optimizing an overall objective function. In other papers [37], [20], [21], [22], [26] and [9] one can find solutions of only very specialized games and first of all for the open-loop case.

The scant research on approximation and numerical methods for differential games and the need to solve many realistic games has given the motivation to undertake this research. In the present chapter we consider a continuous-time, two-person, deterministic differential game with a nonlinear state equation and two quadratic cost functionals.

The nonlinearity of the state equation appears as a regular perturbation term, assumed , to be a polynomial in the state of order higher than one. The presence of the perturbation term determines the use of the classical perturbation method [3], [38]. For the optimal control problems there were several attempts to treat nonlinear analytic systems [12], [1], [24], [27]. In some sense we follow these works and try to approximate players' strategies as truncated power series of a small parameter. We consider the feedback Nash equlibrium strategies, but the proposed method can be extended also to non-inferior and minimax strategies. The necessary conditions for optimality lead to a nonlinear two-point boundary value problem and they are studied in Section 3. Section 4 describes the details of power series, in the small parameter, expansions of the desired Nash equilibrium solution. The solution requires to solve the coupled Riccati matrix differential equations, known from linear-quadratic Nash differential games, and a sequence, up to some order, of quasi-linear partial differential equations.

In Section 5, asymptotic properties of the approximation to the Nash strategies are presented. Several theorems are included. Finally, Section 6 is devoted to remarks on the presented successive approximation.

2. Formulation of the problem

We consider a two-person, nonlinear differential game described by

$$\dot{x} = A(t)x + \varepsilon f(x,t) + B_1(t)u_1 + B_2(t)u_2 \tag{2.1}$$

$$x(0) = x_o \tag{2.2}$$

where

x is an n-vector describing the state of the game,

u_1 is an m_1-vector representing the strategies of Player 1,

u_2 is an m_2-vector representing the strategies of Player 2,

$f(x,t)$ is an n-vector whose elements consist of polynomial in the state variable satisfying $f(0,0)=0$,

ε is a small scalar parameter,

$A(t)$ is an $n \times n$ system matrix-function, piecewise continuous in t,

$B_1(t)$ is an $n \times m_1$ control matrix-function of Player 1, piecewise continuous in t,

$B_2(t)$ is an $n \times m_2$ control matrix-function of Player 2, piecewise continuous in t.

The objective for Player i, i=1,2, is to choose his strategy u_i in such a way to minimize the following performance index

$$J_i(u_1,u_2)=\frac{1}{2}x'(T)Q_{if}x(T)+\frac{1}{2}\int_0^T [x'Q_i(t)x+u_i'R_{ii}(t)u_i+u_j'R_{ij}(t)u_j]dt \qquad (2.3)$$

where j=2 if i=1 and j=1 if i=2, the terminal time T is fixed, and

Q_{if} is an $n\times n$ positive - semidefinite symmetric matrix,

Q_i is an $n\times n$ positive - semidefinite symmetric matrix-function whose elements are piecewise continuous in t,

R_{ij} is an $m_i\times m_j$ positive - semidefinite for $i\neq j$, or positive - definite for $i=j$, symmetric matrix-function with elements piecewise continuous in t.

In a game where coopeartion between players is impossible to be accepted the players are interested in solution which have the Nash equilibrium property [34].

Definition 2.1

A strategy set $\{u_1^*,u_2^*\}$ is called a Nash equilibrium strategy set if

$$J_1(u_1^*,u_2^*) \leq J_1(u_1,u_2^*) \qquad (2.4)$$

$$J_2(u_1^*,u_2^*) \leq J_2(u_1^*,u_2). \qquad (2.5)$$

This solution is secure against any attempt by one player to unilaterlly alter his strategy, since that player can only lose by deviating from his equilibrium strategy. It should be noted that the Nash equilibrium strategy does not give the lowest cost. For instance, if the players follow the Stackelberg strategy then they will gain no less than using the corresponding Nash strategy [34].

The performance index (2.3) represents some physical requiremsnts. Namely, the term $x'(T)Q_{if}x(T)$ is referred to the terminal cost and assures a reasonable low value of the state at the final time T. Since $Q_i \geq 0$, the next term $x'Q_ix$ weights the cost with respect to the state. The terms $u_1'R_{11}u_1$ $(R_{11}>0)$ and $u_2'R_{22}u_2$ $(R_{22}>0)$ penalize the performance indices J_1 and J_2, respectively, for the applied controls. While the term $u_2'R_{12}u_2$ $(R_{12}\geq 0)$ means that Player 2 interferes in the cost of Player 1. In the same way, the term $u_1'R_{21}u_1$ $(R_{21}\geq 0)$ denotes that Player 1 interferes in the cost of Player 2. It is tacitly assumed that any inequality constraints on the controls or state variables may be suitable approximated by penalty functions included in the performance indices (2.3) (cf [31]).

In the sequel we will consider approximation to the feedback controls corresponding to complete measurements of the state vector at every moment during the game. In this case the strategies are functions of the state variables and time, and the Nash equilibrium strategy set is denoted by

$$\{u_1^*(x,t),\ u_2^*(x,t)\} \tag{2.6}$$

3. Necessary conditions for Nash solution

In [35] the necessary conditions for the feedback Nash strategy set (2.6) have been derived by an extension of the variational methods used in the optimal control theory. It seems that one can easily obtain these necessary conditions using the minimum principle.

Let us define the Hamiltonian functions H_1 and H_2 for Player 1 and Player 2, respectively, of the problem (2.1)-(2.5) by

$$H_i = \frac{1}{2}(x'Q_i x + u_i' R_{ii} u_i + u_j' R_{ij} u_j + \\ + p_i'(Ax + \varepsilon f + B_1 u_1 + B_2 u_2). \tag{3.1}$$

The Nash equilibrium strategy set $\{u_1^*, u_2^*\}$ is such a strategy set which minimizes $\{H_1, H_2\}$ assuming that both players obey the Nash rules (2.4) and (2.5). The strategies for two players can be written as follows:

$$u_i^*(x,t) = \arg\min_{u_i \in U_i} H_i(x,t,u_1,u_2,p_i,\varepsilon) = -R_{ii}^{-1} B_i' p_i \tag{3.2}$$

The conjugate state vector $p_i(x,t,\varepsilon)$ satisfies the differential equation

$$\dot{p}_i = -\frac{\partial H_i}{\partial x} - \left(\frac{\partial u_j^*}{\partial x}\right)' \frac{\partial H_i(x,t,u_i^*,u_j^*,p_i,\varepsilon)}{\partial u_j^*} \tag{3.3}$$

$i,j=1,2,\ i \neq j$

with the terminal condition

$$p_i(T) = Q_{if} x(T). \tag{3.4}$$

It should be noted that the second term in (3.3) is absent for the open-loop case (this term is also absent in the optimal control problem). The reason is very simple, for the open-loop problem $\partial u_i^*/\partial x = 0$, $i=1,2$. Whenever the second term in (3.3) is nonzero the open-loop and feedback solutions are different. In general, (3.3) constitutes a set of partial differential equations which can be difficult to be solved.

For the Hamiltonian functions H_1, H_2 described by (3.1) the canonical equations for the feedback problem (3.2) are given by

$$\dot{x} = Ax + \varepsilon f - B_1R_{11}^{-1}B_1'p_1 - B_2R_{22}^{-1}B_2'p_2 \tag{3.5}$$

$$\dot{p}_1 = -Q_1x - A'p_1 - \varepsilon\left(\frac{\partial f}{\partial x}\right)'p_1 + \left(\frac{\partial p_2}{\partial x}\right)'B_2R_{22}^{-1}(B_2'p_1 - R_{12}R_{22}^{-1}B_2'p_2) \tag{3.6}$$

$$\dot{p}_2 = -Q_2x - A'p_2 - \varepsilon\left(\frac{\partial f}{\partial x}\right)'p_2 + \left(\frac{\partial p_1}{\partial x}\right)'B_1R_{11}^{-1}(B_1'p_2 - R_{21}R_{11}^{-1}B_1'p_1). \tag{3.7}$$

The appropriate boundary conditions are

$$\begin{aligned} x(0) &= x_o \\ p_1(T) &= Q_{1f}x(T) \qquad \forall x \\ p_2(T) &= Q_{2f}x(T) \qquad \forall x \end{aligned} \tag{3.8}$$

4. Recurrence relation

Equations (3.5)-(3.8) constitute a nonlinear two-point boundary value problem which looks rether complecated. Due to the partial derivatives $\partial p_2/\partial x$ and $\partial p_1/\partial x$ appearing in (3.6) and (3.7), respectively, it is difficult to apply known methods (e.g. the so-called shooting method, the invariant imbedding method or the quasi-linearization method [23], [25]) for solving nonlinear boundary value problems. For example, in the invariant imbedding method equations (3.6)-(3.7) are treated as characteristic equations of a corresponding partial differential equation, the so-called invariant imbedding equation. In our case this equation is a nonlinear first-order hyperbolic system. In addition, the boundary conditions $p_1(T)$ and $p_2(T)$ hold for any x. It is difficult to place property the integration domain of x without a priori knowledge about the bahaviour of the solution x, p_1, p_2 of the two-point boundary value problem.

Therefore, it seems to be reasonable to use out the specific structure of the considered two-point boundary value problem. Namely, equations (3.5)-(3.7) are regularly perturbed. Just for this reason a method of successive approximation based on the method of perturbation in terms of a small parameter ε is applied. It is assumed that the costate p_i, i=1,2 can be expanded in the form

$$p_i(x,t) = \sum_{k=0}^{\infty} \varepsilon^k p_i^{(k)}(x,t) \tag{4.1}$$

Thereby, the optimal strategy (3.2) of both players are assumed to be analytic in ε for $t \in [0,T]$.

It is assumed that the series (4.1) converges for $t \in [0,T]$ and ε is sufficiently small. The derivative of p_1 and p_2 with respect to time t is possible to be obtained in the form

$$\dot{p}_i = \sum_{k=0}^{\infty} \varepsilon^k \left[\frac{\partial p_i^{(k)}}{\partial x} \dot{x} + \frac{\partial p_i^{(k)}}{\partial t}\right], \quad i=1,2 \tag{4.2}$$

under the assumption that the series (4.1) may be differentiated term-by-term (i.e. $\dot{p}_i$ is uniformly convergent).

Using (3.5) and (4.1) the formula (4.2) can be written as follows

$$\dot{p}_i = \sum_{k=0}^{\infty} \varepsilon^k \left\{\frac{\partial p_i^{(k)}}{\partial x}\left[Ax + \varepsilon f - S_{11} \sum_{k=0}^{\infty} \varepsilon^k p_1^{(k)} - S_{22} \sum_{k=0}^{\infty} \varepsilon^k p_2^{(k)}\right] + \frac{\partial p_i^{(k)}}{\partial t}\right\}, \quad i=1,2 \tag{4.3}$$

where

$$S_{ii} = B_i R_{ii}^{-1} B_i' .$$

On substituting (4.1) and (4.3) into (3.6) and (3.7) and equating powers of the parameter ε we obtain.

For $k=0$

$$\frac{\partial p_i^{(0)}}{\partial t} + \frac{\partial p_i^{(0)}}{\partial x}\left[Ax - S_{11}p_1^{(0)} - S_{22}p_2^{(0)}\right]$$

$$= -Q_i x - A' p_i^{(0)} + \left[\frac{\partial p_j^{(0)}}{\partial x}\right]' B_j R_{jj}^{-1}\left[B_j' p_i^{(0)} - R_{ij} R_{jj}^{-1} B_j' p_j^{(0)}\right]. \tag{4.4}$$

It is easy to notice that (4.4) is a quasi-linear partial differential equation whose characteristic differential equations are the following

$$\frac{dt}{dt} = 1$$

$$\frac{dx}{dt} = Ax - S_{11}p_1^{(0)} - S_{22}p_2^{(0)}$$

$$\frac{dp_i^{(0)}}{dt} = -Q_i x - A' p_i^{(0)}$$

$$+ \left[\frac{\partial p_j^{(0)}}{\partial x}\right]' B_j R_{jj}^{-1}\left[B_j' p_i^{(0)} - R_{ij} R_{jj}^{-1} B_j' p_j^{(0)}\right], \quad i,j=1,2, \quad i \neq j. \tag{4.5}$$

These characteristic equations are the same as the canonical equations (3.5)-(3.7) provided that $\varepsilon=0$, i.a. the two-point boundary value problem for the linear game. It is known [35] that the solution of (4.5) can be expressed in the form

$$p_i^{(0)} = P_i(t)x(t), \tag{4.6}$$

where P_i is a symmetric matrix, given by the solution of the Riccati matrix differential equation (after substituting (4.6) into (4.5))

$$\dot{P}_i=-P_iA-A'P_i-Q_i+P_iS_{ii}P_i+P_iS_{jj}P_j+P_jS_{jj}P_i-P_jS_{ij}P_j,\ P(T)=Q_{if}. \tag{4.7}$$

For $k=1,2,3,\ldots$

$$\frac{\partial p_i^{(k)}}{\partial t} + \frac{\partial p_i^{(k)}}{\partial x}(A-S_{11}P_1-S_{22}P_2)x+(A-S_{11}P_1-S_{22}P_2)'p_i^{(k)} \tag{4.8}$$
$$+\left[\frac{\partial p_j^{(k)}}{\partial x}\right]'(S_{ij}P_j-S_{jj}P_i)x+(S_{ij}P_j-S_{jj}P_i)'p_j^{(k)}=h_i^{(k)},\quad p_i^{(k)}(x,T)=0\quad \forall x$$

where

$$h_i^{(k)}=\begin{cases} -(\frac{\partial f}{\partial x})'P_ix-P_if, & \text{if}\quad k=1\\ -(\frac{\partial f}{\partial x})'p_i^{(k-1)}-\frac{\partial p_i^{(k-1)}}{\partial x}f & \end{cases}$$

$$+\sum_{l=1}^{k-1}\{\frac{\partial p_i^{(l)}}{\partial x}[S_{11}p_1^{(k-l)}+S_{22}p_2^{(k-l)}]+[\frac{\partial p_j^{(l)}}{\partial x}]'[S_{jj}p_i^{(k-l)}-S_{ij}p_j^{(k-l)}]\},$$

$$\text{if}\quad k=2,3,\ldots \tag{4.9}$$

with
$$S_{ij}=B_jR_{jj}^{-1}R_{ij}R_{jj}^{-1}B_j'$$

for $i,j=1,2,\ i\neq j$.

It follows that the nonlinear two-point boundary value problem (3.5)-(3.8) can be solved by solving the coupled Riccati matrix differential equations (4.7) and a series of coupled quasi-linear partial differential equations (4.8).

Now, we will specialize the structure of the perturbation term f, assumed to be a polynomial in x. It is easy to notice that if $f(x,t)$ is a polynomial in x of degree q then $h_i^{(1)}$ in (4.9) is also a polynomial in x of degree q.

Now the following proposition will be established.

Proposition 4.1

If $f(x,t)$ in (2.1) is a polynomial in x of degree q, then the solution $p_i^{(k)}$ of (4.8), for $i=1,2$ and $k=1,2,\ldots$ can be written in the form of a polynomial in x of degree $k(q-1)+1$.

Proof

Let us denote, for $k=1,2,\ldots,$ as follows

$$z^i = p_i^{(k)}$$

$$D = A-S_{11}P_1-S_{22}P_2$$

$$E^j = S_{ij}P_j-S_{jj}P_i$$

$$v^i = h_i^{(k)}, \text{ for } i,j=1,2,\ i\neq j,$$

then (4.8) can be written in the form

$$\frac{\partial z^i}{\partial t}+\frac{\partial z^i}{\partial x}Dx=-D'z^i-\left(\frac{\partial z^j}{\partial x}\right)'E^jx-(E^j)'z^j+v^i \tag{4.10}$$

with $z^i(x,T)=0 \quad \forall x, \quad x(0)=x_o$.

By an easy investigation, we notice that (4.10) can be written in the following form

$$\frac{\partial z^i_\alpha}{\partial t}+\sum_{\beta=1}^{n}a_\beta\frac{\partial z^i_\alpha}{\partial x_\beta}=b^i_\alpha, \quad z^i_\alpha(x,T)=0 \quad \forall x \tag{4.11}$$

where

$$a_\beta=\sum_{\gamma=1}^{n}D_{\beta\gamma}x_\gamma \tag{4.12a}$$

$$b^i_\alpha=v^i_\alpha-\sum_{\gamma=1}^{n}D_{\gamma\alpha}z^i_\gamma-\sum_{\beta=1}^{n}\frac{\partial z^j_\beta}{\partial x_\alpha}\sum_{\gamma=1}^{n}E^j_{\beta\gamma}x_\gamma-\sum_{\gamma=1}^{n}E^j_{\gamma\alpha}z^j_\gamma \tag{4.12b}$$

with $i,j=1,2,\ i\neq j,\ \alpha,\beta=1,2,\ldots,n$ and $D_{\alpha\beta}$ and $E^i_{\alpha\beta}$ are elements of the matrices D and E^i, respectively, while z^i_α are elements of the vector z^i.

The coefficients a_β, $\beta=1,2,\ldots,n$, depend on the variables t and x, while the coefficients b^i_α, $i=1,2$, $\alpha=1,2,\ldots,n$, depend on the variables t,x,z^i and z^j, $j=1,2$, $i\neq j$. The coefficients a_β are identical in all the equations (4.11), for each $\alpha=1,2,\ldots,n$. It is said, that the differential equations of such system have "the same principal part" [5].

Now let the solutions $z^i_1, z^i_2,\ldots,z^i_n$ of (4.11) be given in the implicit form

$$\begin{aligned} \rho_1^i(x,t,z^i,z^j) &= c_1^i \\ \vdots \qquad & \qquad \vdots \\ \rho_n^i(x,t,z^i,z^j) &= c_n^i , \end{aligned} \tag{4.13}$$

these solutions depend on some constant parameters $c_1^i, c_2^i, \ldots, c_n^i$. To assure the possibility of calculating the functions $z_1^i, z_2^i, \ldots, z_n^i$ it is assumed that the determinant of the Jacobian

$$\frac{\partial(\rho_1^i, \rho_2^i, \ldots, \rho_n^i)}{\partial(z_1^i, z_2^i, \ldots, z_n^i)}, \quad i=1,2$$

is different from zero, otherwise the problem has infinitely many solutions.

According to [5, Ch.II, § 2] the integration of the quasi-linear differential equation (4.11) is equivalent to the integration of the system of the following charcteristic differential equations

$$dt = \frac{dx_\beta}{a_\beta} = \frac{dz_\alpha^1}{b_\alpha^1} \tag{4.14a}$$

$$dt = \frac{dx_\beta}{a_\beta} = \frac{dz_\alpha^2}{b_\alpha^2} . \tag{4.14b}$$

Thus the system (4.11) of partial differential equations with the same principal part is equivalent to a system of 3n ordinary differential equations, namely

$$\frac{dx_\beta}{dt} = a_\beta \tag{4.15a}$$

$$\frac{dz_\alpha^1}{dt} = b_\alpha^1 \tag{4.15b}$$

$$\frac{dz_\alpha^2}{dt} = b_\alpha^2 \tag{4.15c}$$

where $\quad \alpha,\beta=1,2,\ldots,n,$

or in the matrix form

$$\dot{x} = Dx \tag{4.16a}$$

$$\dot{z}^1 = -D'z^1-(E^2)'z^2-(\frac{\partial z^2}{\partial x})'E^2x+v^1 \tag{4.16b}$$

$$\dot{z}^2=-D'z^2-(E^1)'z^1-(\frac{\partial z^1}{\partial x})'E^1x+v^2. \tag{4.16c}$$

Since

- (4.16b) is a linear equation; and
- provided that v^i is a polynomial in x

$$v^i_\alpha = \sum_l \delta^i_{\alpha l} x_1^{l_1} x_2^{l_2} x_3^{l_3} \dots x_n^{l_n} \tag{4.17}$$

with $\sum_{k=1}^{n} l_k=r$, where r denotes the order of the polynomial v^1, $\alpha=1,2,\dots,n$; and - assuming that z^2 is a polynomial in x

$$z^2_\alpha = \sum_l \gamma^2_{\alpha l} x_1^{l_1} x_2^{l_2} x_3^{l_3} \dots x_n^{l_n} \tag{4.18}$$

with

$$\sum_{k=1}^{n} l_k=r$$

then the solution of (4.16b) can be also written in the form of a polynomial in x, namely

$$z^1_\alpha = \sum_l \gamma^1_{\alpha l} x_1^{l_1} x_2^{l_2} x_3^{l_3} \dots x_n^{l_n} \tag{4.19}$$

where $\sum_{k=1}^{n} l_k=r$, $\alpha=1,2,\dots,n$.

In the same way equation (4.16c) may be treated and proved that the solution of (4.16c) can be written in the form of a polynomial in x (4.18).

Substituting the derivative of (4.18) with respect to t and (4.8) into (4.17), and next equating coefficients of the same powers one can obtain the following linear differential equation for the coefficients γ^i_l [22]

$$\sigma(\dot{\gamma}) = \begin{bmatrix} -\bigoplus_{k=1}^{r+1} D_k & \bigoplus_{k=1}^{r+1} E^2_k \\ -\bigoplus_{k=1}^{r+1} E^1_k & -\bigoplus_{k=1}^{r+1} D_k \end{bmatrix} \sigma(\gamma) + \sigma(\delta) \tag{4.19}$$

where $D_k \equiv D'$, $E^i_k \equiv (E^i)'$, i=1,2, while $\sigma(\cdot)$ denotes a "stacking operator" which transforms a matrix into a vector, and the symbol $\bigoplus_{k=1}^{r+1}$ denotes the Kronecker sum repeated r+1 times, and r is the order of the polynomial v.

Hence z(x,t) can be obtained by solving a set of differential equations for coefficiients.

For k=1, $h_i^{(1)}$ being a polynomial of degree q implies $p_i^{(1)}$ to be a polynomial of degree q. Next, it can be noticed from (4.9) that for k=2,3,... $h_i^{(k)}$ is a polynomial provided that $p_i^{(l)}$, for $l < k$, is a polynomial. In this way we have shown that $p_i^{(k)}$ can be expressed in the form of a polynomial in x of degree equal to the degree of the polynomial $h_i^{(k)}$.

A short investigation allows us to notice that degree of the polynomial (4.9) is equal to

$$(q-1)+[(k-1)q-(k-2)] = k(q-1)+1 \quad \text{for } k=1,2,3,\ldots \tag{4.20}$$

and the degree of $p_i^{(k)}$ is equal to (4.20). □

Hence the problem of solution of the nonlinear Nash differential game is transformed to the solution of some set of linear ordinary differential equations.

5. Asymptotic properties of the approximation

The solution of (4.8) for k=1,2,... generates a series being the solution of the Nash game. For practical considerations this series has to be truncated and only a few first terms can be taken into account. We will consider the approximate startegy

$$\bar{u}_i(x,t) = -R_{ii}^{-1}B_i'[P_i x + \sum_{k=1}^{w_i} \varepsilon^k p_i^{(k)}(x,t)], \quad i=1,2 \tag{5.1}$$

where w_i is the order of approximation and determines that w_i terms are included in the truncated series of Player i. We will study the approximate strategy $\{\bar{u}_1,\bar{u}_2\}$ as $\varepsilon \to 0$. The results are given in the following theorems.

Theorem 5.1

If P_i is the solution of (4.7) and $p_i^{(k)}$, k=1,2,..., i=1,2, is the solution of (4.8) then the performance index J_i given by (2.3) can be written in the form

$$J_i(x,t) = \frac{1}{2}x'(t)P_i(t)x(t) + \sum_{k=1}^{\infty} \varepsilon^k J_i^{(k)}[x(t),t], \quad i=1,2 \tag{5.2}$$

where $t \in [0,T]$, $J_i^{(k)}$ is the solution of the following equation

$$\frac{\partial J_i^{(k)}}{\partial t} + x'(A-S_{11}P_1-S_{22}P_2)' \frac{\partial J_i^{(k)}}{\partial x} = h_i^{(k)} \tag{5.3}$$

with

$$h_i^{(k)} = \begin{cases} -x'P_i f + x'(P_i S_{jj} - P_j S_{ij}) p_j^{(1)} & \text{if } k=1 \quad (5.4a) \\ -f' \dfrac{\partial J_i^{(k-1)}}{\partial x} + x'(P_i S_{jj} - P_j S_{ij}) p_j^{(k)} & \end{cases}$$

$$+ \sum_{l=1}^{k-1} \{ [S_{11} p_1^{(l)} + S_{22} p_2^{(l)}]' \frac{\partial J_i^{(k-l)}}{\partial x}$$

$$- p_i'^{(l)} S_{ii} p_i^{(k-l)} - p_j'^{(l)} S_{ij} p_j^{(k-l)} \} \tag{5.4b}$$

where

$$S_{ij} = B_j R_{jj}^{-1} R_{ij} R_{jj}^{-1} B_j', \quad i,j=1,2, \quad i \neq j.$$

Proof

For the considered problem (2.1)-(2.3) the Hamilton-Jacobi equation for Player i, i=1,2, [35] is as follows

$$\frac{\partial J_i}{\partial t} + \min_{u_i}\{\frac{1}{2}(x'Q_i x + u_i' R_{ii} u_i + u_j' R_{ij} u_j) + (Ax + \varepsilon f + B_1 u_1 + B_2 u_2)' \frac{\partial J_i}{\partial x}\} = 0 \tag{5.5}$$

for $t \in [0,T]$. The minimized expression in (5.5) will be identical with the expression (3.1), determining the Hamiltonian function H_i, if we replace the gradient of J_i by the vactor p_i. According to (3.2) this expression is minimized by

$$u_i^*(x,t) = -R_{ii}^{-1} B_i' \frac{\partial J_i}{\partial x} = -R_{ii}^{-1} B_i' p_i . \tag{5.6}$$

Substituting

$$u_i(x,t) = -R_{ii}^{-1} B_i' \sum_{k=0}^{\infty} \varepsilon^k p_i^{(k)}$$

into (2.3) we obtain

$$J_i = \frac{1}{2} x'(T) Q_{if} x(T) + \frac{1}{2} \int_0^T x'(Q_i + P_i S_{ii} P_i + P_j S_{ij} P_j) x \, dt$$

$$+ \text{(terms depending of higher degree of } \varepsilon). \tag{5.7}$$

The first line of (5.5) corresponds to the linear model of the game (ε=0) and is equal to [2]

$$J_i^{(0)}(x,t) = \frac{1}{2} x'(t) P_i(t) x(t) \qquad t \in [0,T] \tag{5.8}$$

while the second line of (5.7) consists of a power-series of ε and the coefficients of powers of ε are denoted by $J_i^{(k)}$, $k=1,2,\ldots$.

Substituting (5.2) into (5.5) and equating powers of ε one can easily obtain (5.3)-(5.4).

□

The next theorem shows the relation between $J_i^{(k)}$ and $p_i^{(k)}$, namely:

Theorem 5.2

If $p_i^{(k)}$ *is the solution of* (4.8) *and* $J_i^{(k)}$ *is the solution of* (5.3), *then*

$$\frac{\partial J_i^{(k)}}{\partial x} = p_i^{(k)}, \quad J_i^{(k)}(x,T) = 0 \quad \forall x \tag{5.9}$$

for $k=1,2,\ldots$

The following lemma will be used in the proof of the above theorem.

Lemma 5.3

The equation

$$\frac{\partial v(x,t)}{\partial t} + \frac{\partial}{\partial x}\left[x'R'(t)v(x,t)\right] = 0 \tag{5.10}$$

with $\quad v(x,T) = 0 \quad \forall x$

has a unique solution $v(x,t)=0 \quad \forall x$ *and* $t \in [0,T]$.

Proof

The lemma can be proved in a similar way as Proposition 4.1. The equation (5.10) is written as

$$\frac{\partial v}{\partial t} + \frac{\partial v}{\partial x} Rx = -R'v. \tag{5.11}$$

For each component of the n-vector v we have

$$\frac{\partial v_\alpha}{\partial t} + \sum_{\beta=1}^{n} a_\beta \frac{\partial v_\alpha}{\partial x} = b_\alpha, \quad v_\alpha(x,T)=0 \quad \forall x \tag{5.12}$$

where

$$a_\beta = \sum_{\gamma=1}^{n} R_{\beta\gamma} x_\gamma$$

$$b_\alpha = -\sum_{\gamma=1}^{n} R_{\gamma\alpha} v_\gamma$$

$$\alpha,\beta = 1,2,\ldots,n.$$

The differential equations (5.12) have the same principal part. The system of quasi-linear partial differential equations (5.12) in n+1 independent variables is equivalent to 2n ordinary differential equations

$$dt = \frac{dx_\beta}{a_\beta} = \frac{dv_\alpha}{b_\alpha} \tag{5.13}$$

which are the characteristic differential equations of (5.12). We are interested in the solution of the equations

$$dt = \frac{dv_\alpha}{b_\alpha}, \quad =1,2,\dots,n$$

or written in the matrix form

$$\frac{dv}{dt} = -R'v, \quad v(x,T) = 0 \quad \forall x \tag{5.14}$$

which is as follows

$$v(x,t) = \exp\left[-R(t)\right]c.$$

The constant vector $c=0$ due to the boundary conditions, hence

$$v(x,t) = 0 \quad \forall x \quad \text{and} \quad \forall t.$$

□

<u>Proof of Theorem 5.2</u>

For each $k=1,2,\dots$ we differentiate (5.3) with respect to x. Next, the result we subtract from (4.8) and then we obtain the following equation

$$\frac{\partial}{\partial t}\left[\frac{\partial J_i^{(k)}}{x} - p_i^{(k)}\right] + \frac{\partial}{\partial x}\{x'(A-S_{11}P_1 - S_{22}P_2)'\left[\frac{\partial J_i^{(k)}}{\partial x} - p_i^{(k)}\right]\} = 0 \tag{5.15}$$

with the boundary condition

$$\frac{\partial J_i^{(k)}(x,T)}{\partial x} - p_i^{(k)}(x,T) = 0.$$

Due to Lemma 5.3 it follows that

$$\frac{\partial J_i^{(k)}}{\partial x} = p_i^{(k)}.$$

□

The next theorem concerns the influence of the approximate strategy $\{\bar{u}_1,\bar{u}_2\}$ on the performance indices.

Theorem 5.4

If

$$\bar{p}_i(x,t) = P_i x + \sum_{k=1}^{w_i} \varepsilon^k p_i^{(k)}(x,t) , \quad i=1,2 \tag{5.16}$$

is optimal up to the order w_i in ε, then the performance index J_i is optimal up to the order $w=\min\{2w_1,2w_2\}$ in ε.

Proof

Let $\bar{p}_i$ be a solution of (4.8) up to the order w_i, then the strategy can be written as

$$\bar{u}_i(x,t) = -R_{ii}^{-1}B_i'[P_i x + \sum_{k=1}^{w_i} \varepsilon^k p_i^{(k)}(x,t)], \quad i=1,2. \tag{5.17}$$

Substituting (5.17) into (2.3) we can observe that due to the quadratic terms in (2.3) the performance index J_i is expanded up to the order $w=\min\{2w_1,2w_2\}$ in ε as follows

$$J_i(\bar{u}_1,\bar{u}_2) = J_i(u_1^*,u_2^*) + \varepsilon^w \Lambda_i \tag{2.18}$$

where $\{u_1^*,u_2^*\}$ denotes the optimal strategy set $(w_i=\infty,\ i=1,2)$, while Λ_i is the rest of the power series.

□

The next theorem shows that $\{\bar{u}_1,\bar{u}_2\}$ has the asymptotic Nash property.

Theorem 5.5

The approximate strategy $\{\bar{u}_1,\bar{u}_2\}$ given by (5.17) has the following asymptotic Nash property

$$J_1(\bar{u}_1,\bar{u}_2) \le J_1(u_1,\bar{u}_2) + 0(\varepsilon^{2w_1}) \tag{5.19}$$

$$J_2(\bar{u}_1,\bar{u}_2) \le J_2(\bar{u}_1,u_2) + 0(\varepsilon^{2w_2}) \tag{5.20}$$

$\forall\, u_i \in U_i$, $i=1,2$ of order $2w_2$ and $2w_1$, respectively.

Proof

Let us assume that

$$p_i(x,t) = P_i x + \sum_{k=1}^{\infty} \varepsilon^k p_i^{(k)}(x,t)$$

converges. Theorem 5.4 implies that

$$\left.\frac{\partial^m [J_1(\bar{u}_1,\bar{u}_2) - J_1(u_1^*,u_2^*)]}{\partial \varepsilon^m}\right|_{\varepsilon=0} = 0 \tag{5.21}$$

for $m=0,1,\ldots,2w_1-1$.

By virtue of Theorem 5.4, (5.21) and due to convexity of the performance index we have

$$J_1(\bar{u}_1,\bar{u}_2) = J_1(u_1^*,\bar{u}_2) + 0(\varepsilon^{2w_1})$$
$$\leq J_1(u_1,\bar{u}_2) + 0(\varepsilon^{2w_1}) \qquad \forall u_1 \in U_1$$

which proves (5.19). The proof of (5.20) is similar.

6. Conclusions

In this chapter an approximate solution of a regularly perturbed, nonlinear differential Nash game with quadratic costs is outlined. The nonlinear two-point boundary value problem associated with the necessary conditions for optimality has been replaced by a coupled matrix Riccati differential equation and a series of vactor-value quasi-linear partial differential equations. Provided that the perturbation term is a polynomial in x, the problem of solving the quasi-linear equations is reduced to finding the solution of the corresponding sets of linear ordinary differential equations for the coefficients of the solution polynomial.

The strategy of each player has the form of a power series and there is a region of the state space, for a prescribed value of ε, in which the strategy series is convergent, or for a prescribed region of the state space there is the maximum value of ε for which the strategy series is convergent. From a practical point of view only a few first terms are sufficient to provide adequate strategies. The number of computing terms depends on the value of ε. These problems need a further investigation.

The essential merits of the considered successive approximation of the Nash equilibrium strategies have been demonstrated (via several theorems):

- strategies are in a feedback form
- there is a relation between $J_i^{(k)}$ and $p_i^{(k)}$
- the order of the performance index approximation is twice the order of the strategy approximation
- the approximate strategies have the asymptotic Nash property.

References

[1] J.F. Baldwin, J.H. Sims Williams: The Use of a Method of Perturbations in the Synthesis of Closed-Loop Optimal Control Laws for

Nonlinear Systems, Automatica, 5 (1969), 357-367.

[2] T. Basar, G.J. Olsder: Dynamic Noncooperative Game Theory, Academic Press, New York, 1982.

[3] R. Bellman: Perturbation Techniques in Mathematics, Physics, and Engineering, Holt, New York, 1964.

[4] D.P. Bertsekas: Dynamic Programming and Stochastic Control, Academic Press, New York, 1976.

[5] R. Courant, D. Hilbert: Methods of Mathematical Physics, Interscience Publishers, New York, 1962.

[6] J.B. Cruz, Jr (ed,): Feedback Systems, MeGraw-Hill, New York, 1972.

[7] P.L. Falb, J.L. de Jong: Some Successive Approximation Methods in Control and Oscillation Theory, Academic Press, New York,1969.

[8] N. Farber, J. Shinar: Approximate Solution of Singularly Perturbed Nonlinear Pursuit-Evasion Games, JOTA, 32 (1980), 39-73.

[9] G. Feichtinger, S. Jorgensen: Differential Game Models in Management Science, European J. Oper. Research, 14 (1983), 137-155.

[10] M.H. Foley, W.E. Schnitendorf: On a Class of Nonzero-Sum, Linear -Quadratic Differential Games, JOTA, 7 (1971), 357-377.

[11] B.F. Gardner, Jr. J.B. Cruz, Jr: Well-Posedness of Singularly Perturbed Nash Games, J. Franklin Institute, 306 (1978), 355-374.

[12] W.L. Garrard, N.H. McClamroch, L.G. Clark: An Approach to Suboptimal Feedback Control of Nonlinear Systems, Int. Journal Control, 5 (1967), 425-435.

[13] W.L. Garrard: Additional Results on Sub-optimal Feedback Control of Nonlinear Systems, Int. J. Control, 10 (1969), 657-663.

[14] V. Gourishankar, A. Salama: A Technique for Solving a Class of Differential Games, Int. J. Control, 15 (1972), 529-539.

[15] C.R. Hadlock: Asymptotic Performance of Near-Optimum Controls Obtained by Regular and Singular Perturbations, J. Math. Analysis and Appl., 61 (1977), 292-301.

[16] Y.C. Ho, A.E. Bryson, Jr: Applied Optimal Control, Wiley, New York, 1975.

[17] D. Holt, R. Mukundan: A Nash Algorithm for a Class of Nonzero-sum Differential Games, Int. J. Systems Sci., 2 (1972), 379-387.

[18] H. Khalil: Approximation of Nash Strategies, IEEE Trans. Autom. Control, Ac-25 (1980), 247-250.

[19] P.V. Kokotovic, J.B. Cruz, Jr: An Approximation Theorem for Linear Optimal Regulators, JOTA, 27 (1969), 249-252.

[20] M. Krawczak, K. Mizukami: On the Application of Delayed Sampled Data Onservers to some Interception Problems, Int. J. Systems Sci., 14 (1983), 1099-1113.

[21] M. Krawczak: Regularly Perturbed Zero-Sum Differential Games, Polish-DDR Conference, Warsaw, May 1984.

[22] M. Krawczak: Nash Game with Regular Perturbation, 12th IFIP Conference, Budapest, Sept. 1985.

[23] E.S. Lee: Quasilinearyzation and Invariant Imbedding, Academic Press, New York, 1968.

[24] D.L. Lukes: Optimal Regulation of Nonlinear Dynamical Systems, SIAM J. Control, 7 (1969), 75-100.

[25] G.H. Meyer: Initial Value Methods for Boundary Value Problems, Academic Press, New York, 1973.

[26] K. Mizukami, M. Krawczak: On Heuristic Strategies for some Interception Problem, Control and Cybernetics, 10 (1981), 231-242.

[27] Y. Nishikawa, N. Sannomiya, H. Itakura: A Method for Suboptimal Design of Nonlinear Feedback Systems, Automatica, 7 (1971), 703-712.

[28] G.P. Papavassilopoulos, J.V. Medanic, J.B. Cruz, Jr: On the Existence of Nash Strategies and Solutions to Coupled Riccati Equations in L-Q Games, JOTA, 28 (1979), 49-76.

[29] G.P. Papavassilopoulos, J.B. Cruz, Jr: On the Existence of Solutions to Coupled Matrix Riccati Differential Equations in Linear -Quadratic Nash Games, IEEE Trans. Autom. Control, AC-24 (1979), 127-129.

[30] L.F. Pau: Differential Games and a Nash Equilibrium Searching Algorithm, SIAM J. Control, 13 (1975), 835-852.

[31] E. Polak: Computational Methods in Optimization, Academic Press, New York, 1971.

[32] V.H. Quintana, E.J. Davison: Two Numerical Techniques to Solve Differential Game Problems, Int. J. Control, 16 (1971), 465-474.

[33] S.M. Roberts: Variational Perturbation Method and Power-Series Approximation Method, JOTA, 32 (1980), 441-450.

[34] M. Simaan, J.B. Cruz, Jr: On the Stackelberg Strategy in Nonzero -Sum Games, JOTA, 11 (1973), 533-555.

[35] A.W. Starr, Y.C. Ho: Nonzero-Sum Differential Games, JOTA, 3, (1969), 184-206.

[36] A.W. Starr, Y.C. Ho: Further Properties of Nonzero-Sum Differential Games, JOTA, 3 (1969), 207-219.

[37] D. Tabak: Numerical Solutions of Differential Game Problems, Int. J. Systems Sci., 6 (1975), 591-599.

[38] M. Van Dyke: Perturbation Methods in Fluid Mechanics, Academic Press, New York, 1964.

Chapter 10

QUASI - SUBGRADIENT ALGORITHMS FOR CALCULATING SURROGATE CONSTRAINTS

Jarosław Sikorski

1. Introduction

Calculating an effective bound on the primal optimal value of the candidate problem is of great importance to any branch-and-bound method for solving integer programming problems. In many successful approaches the linear programming relaxation was applied for this purpose. The Lagrangean dual program as a source of closer bounds has attracted attention since aubgradient search methods of solving it were introduced. This led to some effective applications of the Lagrangean duality in solving specially structured problems. Even a more efficient bound is offered by the surrogate dual because it yields usually a tighter duality gap than the Lagrangean dual. However, the solution of the corresponding methematical program is a rather difficult problem for it requires maximizing a quasiconcave, often discontinuous, function.

In this chapter a certain method for calculating surrogate constraints is discussed theoretically and verified numerically in order to show that we can consider it promising for solving the surrogate dual program. Any efficient method of solving it allows us to take advantage of the better bound provided by this approach. An algorithmic scheme of the described method is the central point of the chapter. It is based on the concept of quasi-subgradient which generalizes the subgradient for quasi-concave functions. Hence it is similar to the schemes of subgradient search methods of solving concave maximization problems.

The theoretical framework is presented hereafter for a general formulation of the primal problem so all results are valid in a class of nonconvex mathematical programming problems which includes integer programming problems. Moreover, and this should perhaps be stressed, we use a cone representation of primal constraints and this allows us to discess uniformly both types of them: equality and inequality constraints. Till now, only the latter type was considered when algorithms for calculating surrogate constraints were constructed.

The surrogate constraints as a convex combination of primal constraints was introduced by Glover [10] for binary integer-programming problems. Soon after that Luenberger [22] formulated duality theory in quasi-convex programming. He used, in fact, the same idea of substituting a single constraint, so called surrogate constraint, for a whole set of primal constraints. Basic theoretical properties of the surrogate dual problem which consists in finding best surrogate constraints were described by Glover [11, 12] and Greenberg, Pierskalla [13]. Greenberg [16] considered nonlinear combinations of primal constraints as surrogate constraints. He also proposed to construct two separate surrogate constraints for problems with mixed types of constraints: a single inequality for all inequalities and a single equality for remaining equality constraints. This seemed rather to be a more theoretical than practical proposition since a surrogate problem with two constraints is much less attractive numerically than a problem with only one constraint. This is especially important for integer programming problems. In that class a problem with one constraint is of "knapsack" type and can be solved efficiently using one of many well-known algorithms.

Since maximization of a quasiconcave function is the crucial point of the surrogate dual program this type of functions has attracted much attention. Greenberg and Pierskalla [14] reviewed properties of quasiconvex functions. The same authors [15] presented a theory of quasiconjugates following the theory of conjugates for convex functions [25]. The advanced results of Crouzeix [2] and Oettli [23] remain in the same stream of theoretical works concerning duals for general quasiconvex primal problems.

Relationships between surrogate and Lagrangean duals were discussed by Karwan and Rardin [18] for linear integer programming problems and by Dyer [14] for a broader class of convex problems. In the latter paper two algorithms for calculating surrogate constraints are presented. Other algorithms were intrudced by Glover [10], Karwan and Rardin [19], Karwan, Rardin and Sarin [20, 21]. Two successful applications of surrogate duality in solving specially structured integer programming problems were reported by Bricker [1] and Fisher, Lageweg, Lenstra, Rinooy kan [7]. Dinkel and Kochenberger [3] utilized the surrogate dual in geometrical programming.

The main part of the chapter consists of four sections. In the first one a theoretical framework of surrogate duality is sketched for a nonconvex mathematical programming problem with a cone representation of constraints. In the second an algorithmic scheme is described and the main convergence result is proved. The third section contains a more detailed description of algorithms for calculating surrogate constraints in the cases where primal problem has different types of constraints. Results of numerical experiments are reported in the fourth one. The paper is provided with Appendix containing data for the test problems.

2. Basic theoretical results

We state the primal problem as

$$\text{(P)} \qquad \hat{f} = \min\{f(x) : x \in X\}, X = \{x \in X_o : -g(x) \in D\}$$

where $f : R^n \to R$, $g : R^n \to R^m$, $X_o \subset R^n$, $D \subset R^m$.

We assume that: $X=\emptyset$, f is lower semicontinuous and g is continuous on X_o, X_o is compact, D is a convex closed cone with its vertex at $0 \in R^m$. Under these assumptions X is compact so $\hat{f}$ is attained by f. In particular, X_o can be a discret set so (P) can represent integer programming problems too. If $D=R^m_+$ (R^m_+ denotes the nonnegative orthant in R^m) then (P) contains inequality constraints but if $D=\{0\}$ then only equality constraints appear in it.

The approach presented here consists in replacing the primal problem by a relaxed one which can be solved more easily. With this purpose we extend the feasible set X be defining a constraint condition weaker than $-g(x) \in D$. Let $D^* = \{\omega \in R^m : \langle u,\omega \rangle \geqslant 0$ for all $u \in D\}$ be a dual cone for D ($\langle .,. \rangle$ denotes scalar product on $R^m \times R^m$). D^* is closed and convex. For any $\omega \in D^*$ we define the set $S(\omega) = \{u \in R^m : \langle u,\omega \rangle \geqslant 0\}$ that is a closed halfspace for $\omega \neq 0 \in R^m$ ($S(0)=R^m$). After having defined $S(\omega)$ we can introduce the following surrogate problem

$$\text{(S)} \qquad h(\omega) = \min\{f(x) : x \in X(\omega)\}, \; X(\omega) = \{x \in X_o : -g(x) \in S(\omega)\}.$$

Thus the cone D has been replaced by $S(\omega)$ which contains it. Moreover, $D = \bigcap_{\omega \in D^*} S(\omega)$. The function $h : D^* \to R$ is well defined because all sets $X(\omega)$ are compact and nonempty. We shall cal h the surrogate dual function. An equivalent definition of $X(\omega) : \{x \in X_o : \langle g(x),\omega \rangle \leqslant 0\}$ shows a single inequality called the surrogate constraint. Notice that the surrogate constraint has the form of inequality regardless of the type of primal constraints. This is in accord with Lagrangean duality in which constraints of both types are included in a modified objective function in the same way. The only difference appears in the dual feasible set where Lagrangean dual multipliers corresponding to the inequality constraints must be nonnegative. For that reason elements of D^* which parametrize (S) will be called the surrogate dual multipliers.

The inclusion $X \subset X(\omega)$ for all $\omega \in D^*$ implies that $h(\omega) \leqslant \hat{f}$. Thus the surrogate dual problem aimed at finding the best surrogate constraint (if it exists) is stated as

(SD) $$\hat{h} = \sup \{h(\omega) : \omega \in D^*\}$$

We recall, for comparison, that the Lagrangean dual approach consists in including primal constraints into an objective function by means of multipliers also from D^*. The Lagrangean dual function is defined by the solution of the following relaxed problems:

(L) $$L(\omega) = \min \{f(x) + \langle g(x), \omega \rangle : x \in X_o\}$$

The Lagrangean dual program is stated as

(LD) $$\hat{L} = \sup \{L(\omega) : \omega \in D^*\}$$

All relationships between (SD) and (LD) are consequences of the fact that functions $\langle g(\cdot), \omega \rangle : R^n \to R$, for $\omega \in D^*$, play a crucial role in both approaches and they are used differently.

It should be mentioned that for all parts of presentation where (LD) is discussed the asumption of lower semicontinuity of f ought to be strengthened to continuity of it.

The following two properties of the set $X(\omega)$ in (S): $X(0) = X_o$ and $X(\alpha\omega) = X(\omega)$ for $\alpha > 0$, yield respectively two features of the surrogate dual function: $h(0) \leqslant h(\omega)$ for all $\omega \in D^*$ and $h(\alpha\omega) = h(\omega)$. Thus zero may be excluded from the feasible set in (S) and an arbitrary normalisation of ω's is possible which allows us to restrict the feasible set in (SD) to a compact one. This opportunity is exploited for all algorithms presented in Section 4 but we do not make use of it in this section.

All results presented below strictly correspond to the results reported in [11, 13, 12] and [4] in the case of inequality constraints ($D = R^m_+$). They can be proved easily for the present formulation of the primal problem by a straightforward modification of the existing proofs. Hence many proofs in this section are omitted and they appear only if the result is more general or its correspondence to a known one is less clear.

<u>Proposition 2.1</u> [11] (Weak Duality Theorem)

$$\hat{h} \leqslant \hat{f}$$

Proposition 2.2 [13]

$\hat{h}$ is achieved and $\hat{h}=\hat{f}$ if and only if $h(\bar{\omega})=f(\bar{x})$ for some $\bar{\omega} \in D^*$ and $\bar{x} \in X$. Then $\bar{x}$ solves (P) and $\bar{\omega}$ solves (SD).

If the equality does not hold the quantity $(\hat{f}-\hat{h})$ is called the surrogate duality gap. The smaller the duality gap, the more valuable is the problem (SD). Even if there is no duality gap $\hat{h}$ may not be attained on D^* by the function h. Such a situation was illustrated in [4] and then the gap was called infinitesimal.

Proposition 2.3 [13]

h is lower semicontinuous and quasiconcave on D^*.

The above result states two basic properties of the surrogate dual function. Quasiconcavity of h appears as a mere consequence of the formulation of (S) and is independent of any additional assumptions which can be imposed on the primal problem. Even if (P) is convex i.e. X_o is convex, f is convex on X_o and g is D-convex, h may still remain quasiconcave but not concave (see example below).

Example 2.1 (n=m=2)

$f(x)=2+x_2-x_1$, $g_1(x)=x_1-x_2$, $g_2(x)=x_1+x_2-2$, $D=R^2_+$, X_o is a semicircle as it is shown in Fig.1. The set X is dashed.

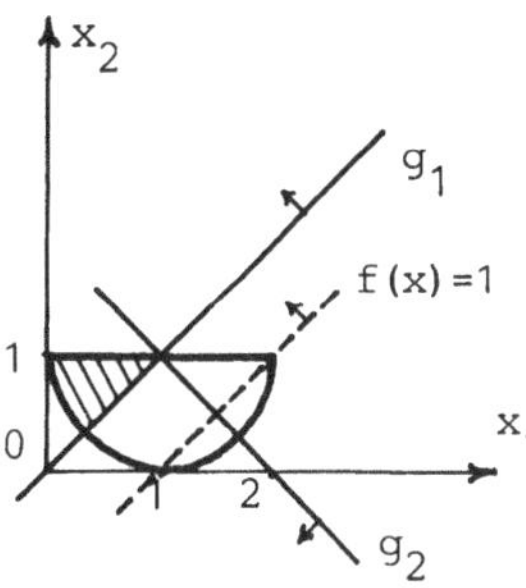

Fig. 1

$h(\omega) = 2 - \frac{\omega_2\sqrt{2}}{\sqrt{\omega_1^2+\omega_2^2}}$ is not concave and it appears in Fig.2 ,

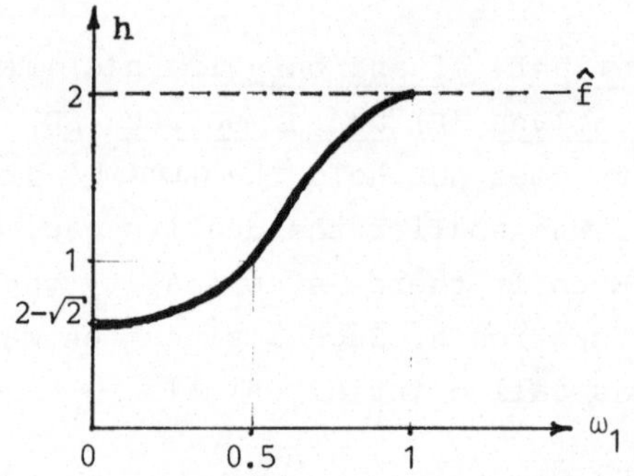

Fig. 2

where the graph of $h(\omega_1)$ is sketched for $\omega_2=1-\omega_1$.

The properties of h given in Proposition 2.3 imply

Corollary 2.1

For any $q<\hat{h}$ the set $W(q)=\{\omega \in D^* : h(\omega)>q\}$ is a nonempty convex cone which is relatively open (i.e. it can be represented as $D^* \cap U$, where U is an open cone).

Continuity of h cannot be assured by assuming that f is continouos. Thus in order to guarantee that $\hat{h}$ is attained on D^* the class of problems considered as (P) must be restricted more radically.

Proposition 2.4 [4]

If the set $F=\{f(x): x \in X_o\}$ has no cluster point then $\hat{h}$ is attained on a relatively open convex cone contained in D^*.

Pure integer programming problems with finite set F satisfy assumptions of Proposition 2.4 so there always exists $\omega \in D^*$ such that $h(\omega)=\hat{h}$ in their duals. Moreover, this equality holds in a neighbourhood of any optimal multiplier . This suggests that the surrogate dual can be more useful for problems of the mentioned class. Thus in that case "sup" in the statement of (SD) can be replaced by "max" and if there is no duality gap then the infinitesimal gap cannot appear.

The problem (SD) is a maximization problem with a quasiconcave lower semicontinuous objective function and with a feasible set that is a convex and closed cone. Hence it is a rather difficult problem from the operative point of view. Some additional characterization of the value $\hat{h}$, different from its definition in (SD), can be helpful. In order to achieve a more geometrical characterization we introduce certain families of sets. Let $T(q)=\{x \in X : f(x) \leqslant q\}$, for any $q \in R$, denote a lower level set of f. Then the inequality $h(\omega) \leq q$ is equivalent to the condition $T(q) \cap X(\omega) \neq \emptyset$. Let $Q(\omega) = \{x \in X(\omega) : f(x) = h(\omega)\}$ be the optimal set of (S). Then clearly $Q(\omega)$ can be expressed as $T(h(\omega)) \cap X(\omega)$. Let $R(q) = g(T(q))$ denote the image of the level set T under the constraint function g used in (P). Define $R^O(q) = \{\omega \in D^* : \langle u,\omega \rangle > 0$ for all $u \in R(q)\}$. Notice that if $0 \in \text{co}R(q)$ then $R^O(q) = \emptyset$ ($\text{co}R(q)$ denotes the convex hull of $R(q)$).

Proposition 2.5 [4]

$$\hat{h} = \min\{q \in R : R^O(q) = \emptyset\}$$

Proposition 2.6

$R^O(q) = \emptyset$ if and only if $\text{co}R(q) \cap -D \neq \emptyset$

Proof

The condition $\text{co}R(q) \cap -D \neq \emptyset$ implies that there exists $u \in \text{co}R(q)$ such that $\langle u,\omega \rangle \leq 0$ for all $\omega \in D^*$. Thus there is no ω in D^* such that $\langle u,\omega \rangle > 0$ for all $u \in R(q)$. This means $R^O(q) = \emptyset$.

$T(q)$ is compact for any $q \in R$ so $R(q)$ is compact since g is continuous. Suppose on the contrary that $\text{co}R(q) \cap -D = \emptyset$. Then the separating hyperplane theorem implies that there exists $\omega \in D^*$ such that $\langle u,\omega \rangle > 0$ for all $u \in \text{co}R(q)$. This means $R^O(q) \neq \emptyset$. □

Corollary 2.2

$$\hat{h} = \min\{q \in R : \text{co}R(q) \cap -D \neq \emptyset\}$$

Stop tests in the algorithms presented in Section 4 are based on the second characterization stated in the above corollary.

In general, dual problems are solved in order to calculate effectively lower bounds on $\hat{f}$: $\hat{L} \leq \hat{f}$ and $\hat{h} \leq \hat{f}$ for example. Usually it is impossible to calculate $\hat{f}$ exactly without solving (P). (Assumptions imposed on (P) in this section are not strong enough). Thus a comparison of the functions $h(\omega)$ and $L(\omega)$ can indicate how useful can be the surrogate dual in the process of bounding $\hat{f}$.

Proposition 2.7

$h(\omega) \geq L(\omega)$ for all $\omega \in D^*$

Proof

It is enough to consider the Lagrangean dual problem for (S) i.e.

$\sup\{L(u):u\in S^*(\omega)\}$, where $L(u)=\min\{f(x)+\langle g(x),u\rangle : x\in X_o\}$ and $S^*(\omega) = \{u:u=\alpha\omega,\ \alpha\geq 0\}$. The Weak Duality Theorem for that dual implies that $h(\omega)\geq\sup\{L(u):u\in S^*(\omega)\}=\sup\{L(\alpha\omega):\alpha\geq 0\}$. This means that $h(\omega)\geq L(\omega)$ for all $\omega\in D^*$ since $h(\alpha\omega)=h(\omega)\ (\alpha>0)$ and clearly $h(0)=L(0)$.

□

Corollary 2.3

If $v\in D^*$ is an optimal Lagrange multiplier then $h(v)\geq\hat{L}=L(v)$.

Corollary 2.4

$$\hat{h}\geq\hat{L}.$$

The relationship between $h(\omega)$ and $L(\omega)$ can be described more precisely and this leads to the conclusion that any optimal Lagrange multiplier can be useful for the solution of the surrogate dual problem (SD) as a first approaximation.

Proposition 2.8

One of the following cases takes place for $\omega\in D^*$:

(i) $h(\omega)>L(\omega)$,

(ii) $h(\omega)=L(\omega)<\hat{h}$ but any neighbourhood of ω contains $u\in D^*\backslash\{0\}$ such that $h(u)>h(\omega)$,

(iii) $h(\omega)=L(\omega)=\hat{h}$ (i.e. ω is an optimal Lagrange multiplier as well as an optimal surrogate multiplier).

Proof

In general, Proposition 2.7 states that $h(\omega)\geq L(\omega)$ and the inequality $f(x)+\langle g(x),\omega\rangle\geq L(\omega)$ for any $x\in X_o$ is a consequence of the statement of (LD).

Assume that $h(\omega)=L(\omega)$ which means that (i) does not hold. Hence $f(x)=L(\omega)$ for all $x\in Q(\omega)$. This implies the inequality $f(x)+\langle g(x),\omega\rangle \leq L(\omega)$ because $\langle g(x),\omega\rangle\leq 0$ (the surrogate constraint is satisfied). In conjunction with the previous, opposite, inequality we have $f(x)+\langle g(x),\omega\rangle=L(\omega)$ and $\langle g(x),\omega\rangle=0$ for all $x\in Q(\omega)$. This is true in both cases (ii) and (iii).

Assume now that (i) does not hold and neither does (iii). Then $\bar{u}\in D^*\backslash\{0\}$ exists such that $h(\bar{u})>h(\omega)$. Let $u^s=(1-s)\omega+s\bar{u}\in D^*\backslash\{0\}$ for $s\in(0;1]$. If $s\to 0$ then $u^s\to\omega$. $h(u^s)\geq h(\omega)$ because h is quasiconcave. For any $x\in Q(\omega)$ the equality $\langle g(x),u^s\rangle=s\langle g(x),\bar{u}\rangle$ holds. If $x\in Q(\omega)$ then $\langle g(x),\bar{u}\rangle>0$ i.e. $x\notin X(\bar{u})$. Hence this equality implies that $x\notin X(u^s)$ either. Thus the intersection $Q(\omega)\cap X(u^s)$ is empty for all $s \in (0;1]$.

We should like to show now that the above condition implies the situation described as (ii).

Suppose on the contrary that $h(u)=h(\omega)$ for all u in a certain neighbourhood of ω. This means $s' \in (0;1]$ exists such that $h(u^{s'})=h(\omega)$. Consequently $x^{s'} \in X(u^{s'})$ exists such that $f(x^{s'})=h(\omega)$. On the other hand, $x^{s'} \notin X(\bar{u})$ i.e. $\langle g(x^{s'}),\bar{u}\rangle > 0$. Hence $(1-s')\ \langle g(x^{s'}),\omega\rangle=\langle g(x^{s'}),u^{s'}\rangle - s'\langle g(x^{s'}),\bar{u}\rangle \leqslant 0$ that means $x^{s'} \in X(\omega)$. Thus $x^{s'} \in Q(\omega) \cap X(u^{s'})$ and the contradiction appears.

Hence the inequality $h(u^{s})>h(\omega)$ holds for all $s \in (0,1]$ so we have (ii). To complete the proof notice that if (iii) holds then $\sup\{L(u): u \in D^*\} \geqslant L(\omega)=\hat{h}$ and Corollary 2.4 implies $L(\omega)=\hat{L}=\hat{h}$.

□

In the case $h(\omega)=L(\omega)$ the local property of the surrogate dual function (ii) is a straight consequence of concavity of the Lagrangean dual function which bounds $h(\omega)$ from below on D^*. If the equality holds at a certain point not being an optimal surrogate multiplier then h does not attain its local maximum at that point.

<u>Corollary 2.5</u> [4]

<u>If</u> v <u>is an optimal Lagrange multiplier then one of the following cases takes place</u>:

(i) $h(v) > \hat{L}$,

(ii) $h(v)=\hat{L}$ <u>but any neighbourhood of</u> v <u>contains</u> $u \in D^*\setminus\{0\}$ <u>such that</u> $h(u) > \hat{L}$,

(iii) $h(v) = \hat{h} = \hat{L}$.

We can consider an optimal Lagrange multiplier to be a good starting point in the process of seeking optimal surrogate dual multipliers. It must be remembered however that its quality is described by comparison of dual function values and not by distance from the set of optimal or nearly optimal surrogate multipliers.

It is known [5] that the Lagrangean dual function is concave on D^* without reference to the convexity of (P). In this section it is shown that the surrogate dual function may be quasiconcave even if (P) is convex. Thus, in general, the upper level sets are convex for both functions but the set $\{(\omega,q) \in R^m \times R : q \leqslant h(\omega)\}$ is not convex while the analogical set for $L(\omega)$ is convex. This implies that the subgradient [25] does not exist everywhere for $h(\omega)$ and cannot be used in analysis as it is extensively used in the case of Lagrangean duality. Moreover $L(\omega)$ is continuous while $h(\omega)$ is only lower semicontinuous.

This means that the lower level sets of the objective functions in (LD) and (SD) are closed but the upper level sets have this property only in the case of (LD).

If the problem (S) is convex i.e. f is convex, g is $S(\omega)$ - convex and X_o is convex, then there is no Lagrangean duality gap for it. Notice that if g is D-convex then it is $S(\omega)$-convex too ($\omega \in D^*\setminus\{0\}$). The opposite implication is not true. Thus (S) can be convex for a certain $\omega \in D^*\setminus\{0\}$ even if (P) is not. Hence the situation $\hat{f} > \hat{h} = \hat{L}$ can also appear. As it is implied by Corollary 2.4 the double equality $\hat{f}=\hat{h}=\hat{L}$ holds always when $\hat{f}=\hat{L}$ (for example when (P) is convex). The situation $\hat{f}=\hat{h} > \hat{L}$ is considered below.

We assume additionally for the further presentation that f is quasiconvex and upper semicontinuous on X_o, g is D-convex and X_o is convex. Such primal problems are called in [22] quasiconvex. Unfortunately, integer programming problems in which X_o is a part of integer lattice or mathematical programming problems with nonlinear equality constraints are not of that type.

Proposition 2.9 [22]

If (P) is quasiconvex and the condition: int $D \neq \emptyset$ and $x \in X_o$ exists such that $-g(x) \in$ int D, is satisfied, then $\bar{\omega} \in D^*\setminus\{0\}$ exists such that any solution of (P) is a solution of (S).

Proof

If two sets are defined: $A=\{a \in R^m : x \in X_o$ exists such that $a-g(x) \in D\}$ and $B=\{a \in A : x \in X_o$ exists such that $a-g(x) \in D$ and $f(x) < \hat{f}\}$ then reasoning can proceed analogously to [22]. The separating theorem is applied for B and -D. The hyperplane is defined by $\bar{\omega} \in D^*\setminus\{0\}$ which existence is a matter of the theorem.

□

Corollary 2.6 (Strong Duality Theorem)

If assumptions of Proposition 2.9 are satisfied then $\hat{h}$ is attained by the surrogate dual function on $D^*\setminus\{0\}$ and $\hat{h}=\hat{f}$.

Thus there is no surrogate duality gap and the case of infinitesimal gap is also excluded for quasiconvex problems satisfying the additional condition of nonempty interior. Hence $\hat{f}$ can be calculated exactly by solving (SD). In general, for such problems the Lagrangean duality gap exists and $\hat{f}$ cannot be calculated but only bounded by solving (LD). However, the set of all assumptions imposed in the primal problem in Proposition 2.9 restricts strongly the class of problems which can be represented by (P). Neither integer programming problems nor equality-constrained problems are included. Since we are interested mainly in these problems Corollary 2.6 states a theoretical result which indicates only how strong assumptions must be made to assure that $\hat{h}=\hat{f}$. In fact, they are also sufficient for continuity of the dual surrogate function [22].

A few words should be said about possible extensions of the duality theorem presented above for quasiconvex problems. The initial assumption that X_o is compact plays important role in this section. It allows us to assure easily that the primal objective function attains its finite minima in all the problems (S). This is fairly convenient for the numerical solution of the surrogate problems. However, this assumptions can be completely neglected and Proposition 2.9 can be proved as well for inequality-constrained problems [2]. A generalized version of that result is presented in [23] for primal problems stated in topological vector spaces where constraints are represented by a multifunction. Then a certain linear functional plays the role of surrogate multiplier and a condition defined for an adequate marginal function is interpreted as a surrogate constraint.

3. A general scheme of maximization

Let us consider the following mathematical program which can represent the versions of (SD) discussed in the next section of this chapter:

$$\text{(MP)} \qquad \hat{h} = \sup\{h(\omega) : \omega \in G,\ p(\omega) \geqslant 0\}$$

Clearly it must be assumed that $h: R^m \to R$ is quasiconcave and lower semicontinuous. We assume additionally that $G \subset R^m$ is convex and closed while $p: R^m \to R^k$ is R^k_+-concave. Moreover, the interior condition must be satisfied for p: there exists $\omega \in G$ such that $p(\omega) > 0$ (i.e. $p(\omega) \in \operatorname{int} R^k_+$).

A recursive scheme which can be useful for the approximative solution of (MP) ought to generate a sequence $\{\omega^i\} \subset G$ that has the following property: for any $\varepsilon > 0$ i_o exists such that $h(\omega^{i_o}) > \hat{h} - \varepsilon$. It is of use to have two search directions at each point of the sequence: one to look for bigger values of h and another one to seek feasible points subject to the condition $p(\omega) \geqslant 0$.

In order to construct the first of the two mentioned above directions a quasi-subgradient of h at ω^i [15] can be used. Transforming to some extent the idea presented in [15] the z-quasiconjugate h_z^* can be defined for the function h.

Definition 3.1

$h_z^*(y)=z - \sup\{h(u) : \langle u,y\rangle \leqslant z\}$.

$h_z^* : R^m \to \bar{R}$ is quasiconcave for any $z \in R$ (it is denoted formally that $\sup\{h(u) : u \in \emptyset\} = -\infty$). It is known [25] that subgradient is an element satisfying as an equality the Fenchel inequality which ties together a convex function with its conjugate. A similar statement containing z-quasiconjugate can serve as a definition of the quasi-subgradient.

Definition 3.2

$y^* \in R^m$ is said to be a quasi-subgradient of h at ω if and only if $h(\omega) + h^*_{\langle y^*,\omega\rangle}(y^*)=\langle y^*,\omega\rangle$.

Existence of at least one quasi-subgradient of h at each point $\omega \in G$ is of fundamental importance for further applications of this element. Instead of formulating here the appropriate theorem in general case we present at the end of this section a result that answers this equestion in the case of the surrogate dual problem (Proposition 3.2). We assume for further presentation of the maximization scheme that it is possible to calculate at least one quasi-subgradient of the function h at each point $\omega \in G$.

The quasi-subgradient has a property which corresponds to the relation tying a subgradient with a maximal point of h.

Lemma 3.1

$0 \in R^m$ is a quasi-subgradient of h at $\bar{\omega} \in R^m$ if and only if $\bar{\omega} = \arg\max h(u)$.

Proof

That $0 \in R^m$ is a quasi-subgradient is equivalent to the equality $h(\bar{\omega}) + h_0^*(0) = 0$ which, after reformulation, takes on the form $h(\bar{\omega}) - \sup\{h(u) : u \in R^m\}=0$.

□

If a subgradient of h exists at ω (denote it by $\bar{y}$) then a supporting functional for the upper level set$\{u : h(u) \geqslant h(\omega)\}$ is generated. Hence the following implication is satisfied by $\bar{y}$ for all $u \in R^m$

(IS) $$h(u) \geqslant h(\omega) \Longrightarrow \langle \bar{y},u\rangle \geqslant \langle \bar{y},\omega\rangle$$

In general, a subgradient may not exist at some point of the domain of quasiconcave function. Obviously there may exist a nonzero element which satisfies (IS) at this point. However, for such a function, there are also points such that only the element $\bar{y}=0$ can satisfy this implication but the function does not attain its global maximum at them. This explains why the concept of quasi-subgradient has been introduced.

In order to compare a property of quasi-subgradient with (IS) we rewrite the equality in Definition 3.2 in an equivalent form using the definition of z-quasiconjugate. This leads to the equality $h(\omega) = \sup\{h(u) : \langle y^*,u\rangle \leqslant \langle y^*,\omega\rangle\}$ which can be expressed in the following form:

(IQ) $\quad \langle y^*,u\rangle \leqslant \langle y^*,\omega\rangle \Longrightarrow h(u) \leqslant h(\omega) \quad$ (for all $u \in R^m$)

The above implication can be interpreted in the same way as (IS) that the quasi-subgradient y^* generates a strongly supporting functional for the interior of the upper level set $\{u : h(u) > h(\omega)\}$. This is not entirely precise but shows the main difference between quasi-subgradient and subgradient. In the case of quasiconcave function a point ω may not belong to the boundary of the set $\{u : h(u) \geqslant h(\omega)\}$ and then a subgradient may not exist at this point.

If an element $\bar{y}$ is a subgradient of h at some point then $\bar{y}$ is also a quasi-subgradient of h at that point [15]. In the case of concave function the set of all quasi-subgradients at any point of its domain is identical with the cone generated by the subdifferential at that point (this results from properties of concave functions [25]).

In general, for a quasiconcave function, there are some elements satisfying (IQ) at a given point which does not satisfy (IS) at the same point. The implication (IQ) indicates that the quasi-subgradient can be used to construct a search direction for maximization of h.

If some point in the sequence generated by a scheme of maximization is infeasible subject to p (i.e. $p(\omega) < 0$) then we need another direction to find the feasible set. The following element is proposed to serve in that case.

<u>Definition 3.3</u>

<u>$p^o \in R^m$ is said to be a supporting element of the function p at $\omega \in R^m$ with respect to R^k_+ if and only if the following implication holds for all $u \in R^m$:</u>

(IP) $\quad p(u) \geqslant P_-(p(\omega)) \Longrightarrow \langle p^o,u\rangle \geqslant \langle p^o,\omega\rangle$

The first inequality in (IP) represents the relation generated by R^k_+. P_- denotes the orthogonal projection on $-R^k_+$ that means $P_-(d) = \bar{d}$, where $\bar{d}_j = \min(0, d_j)$ for $j=1,\dots,k$. We define p^o in order to obtain a direction which can lead to the set $\{u : p(u) \geq 0\}$ if we start from a point $\omega \in R^m$ lying outside of this set. Therefore $p^o = 0 \in R^m$ if $p(\omega) \in \text{int } R^k_+$, since in this case all points in a neighbourhood of ω are feasible. Clearly the inclusion $\{u : p(u) \geq 0\} \subset \{u : p(u) \geq P_-(p(\omega))\}$ holds for all $\omega \in R^m$.

If for all concave functions p_j $(j=1,\dots,k)$ subgradients $\bar{p}_j$ can be calculated at the point ω such that $p(\omega) \notin \text{int } R^k_+$ then p^o can be easily obtained.

Lemma 3.2

If $J = \{j : p_j(\omega) \leq 0\}$ is nonempty then any combination of the form $\sum_{j \in J} \alpha_j \bar{p}_j$, where $\alpha_j \geq 0$ for $j \in J$, is a supporting element of p at ω with respect to R^k_+.

Proof

Consider $u \in R^m$ such that $p(u) \geq P_-(p(\omega))$. The inequality $p_j(u) \geq p_j(\omega)$ holds for $j \in J$. Thus the result is a consequence of implications (IS) holding for p_j and $\bar{p}_j$ $(j \in J)$. □

Usually a subgradient of the function $r(u) = \min\{p_j(u) : j=1,\dots,k\}$. $(r : R^m \to R)$ is used for seeking the feasible set $\{u : p(u) \geq 0\}$ [24]. The function r is concave and for any subgradient $\bar{r}$ at ω the implication of (IS) type holds for all $u \in R^m$

(IR) $$r(u) \geq r(\omega) \Longrightarrow \langle \bar{r}, u \rangle \geq \langle \bar{r}, \omega \rangle$$

Any element satisfying (IR) at $\omega \in R^m$ such that $p(\omega) \notin \text{int } R^k_+$ satisfies (IP) too, since $\{u : p(u) \geq P_-(p(\omega))\} \subset \{u : r(u) \geq r(\omega)\}$. Thus any subgradient $\bar{r}$ at ω is a supporting element of p with respect to R^k_+ but there can exist also such elements p^o which are not subgradients of r.

It is known [24] that if the functions p_j are differentiable (i.e. the only subgradient $\bar{p}_j = \nabla p_j$) then any subgradient of r is of the form $\bar{r} = \sum_{j \in \bar{J}} \alpha_j \bar{p}_j$, where $\bar{J} = \{j : p_j(\omega) = r(\omega)\}$, $\sum_{j \in \bar{J}} \alpha_j = 1$ and $\alpha_j \geq 0$ for $j \in \bar{J}$. It is easy to notice that $\bar{J} \subset J$ for ω such that $p(\omega) \notin \text{int } R^k_+$. Thus using the implication (IP) we have the opportunity to choose a search direction from a wider cone. This allows us, in fact, to interprete properly the algorithms presented in the next section.

Having defined the two search directions we can formulate the recursive scheme designed to approximate the optimal value of (MP). We follow the idea from [24]

$$\text{(RS)} \qquad \omega^{i+1} = P_G(\omega^i + t_i d^i / \|d^i\|) , \ \omega^1 \in G$$

where

$$d^i = \begin{cases} \text{a quasi-subgradient of } h \text{ at } \omega^i, \text{ if } p(\omega^i) \geq 0 \\ \text{a supporting element of } p \text{ at } \omega^i \text{ with respect to } R^k_+, \text{elsewhere} \end{cases}$$

$\{t_i\}$ is a step size sequence which satisfies: $0 < t_i \to 0$ for $i \to \infty$ and $\sum_{i=1}^{\infty} t_i = \infty$, P_G is the orthogonal projection on G, $\|\cdot\|$ denotes the Euclidean norm in R^m.

A quasi-subgradient of the function h is used to construct the search direction for seeking points of bigger objective function value if the current point of the sequence is feasible: $p(\omega^i) \geq 0$. If not then searching concentrates on the feasible set and a supporting element of p is used. The sequence generated by (RS) satisfies the condition $\{\omega^i\}_1^\infty \subset G$.

<u>Proposition 3.1</u>

<u>If all assumptions imposed on</u> (MP) <u>are satisfied then any sequence generated by the scheme</u> (RS) <u>contains a subsequence</u> $\{\omega^{i_s}\}$ <u>such that</u> $\lim_{s\to\infty} h(\omega^{i_s}) = \hat{h}$.

<u>Proof</u>

(RS) is defined for $\omega^i \in G$ such that $h(\omega^i) \neq \hat{h}$. Choose any $q < \hat{h}$ and denote the set $\{\omega : p(\omega) \geq 0, \ h(\omega) \geq q\}$ by $W_p(q)$. There exists $\tilde{\omega} \in G \cap \text{int } W_p(q)$. Set $\delta > 0$ such that $W_p(q)$ contains the closed ball $B(\tilde{\omega};\delta)$.

Let us suppose that $\omega^i \notin W_p(q)$ for all i. If $p(\omega^i) \geq 0$ then $W_p(q) \subset \{\omega : h(\omega) > h(\omega^i)\}$ because $h(\omega^i) < q \leq h(\omega)$ for all $\omega \in W_p(q)$. Then (IQ) implies that $\langle d^i, \omega \rangle > \langle d^i, \omega^i \rangle$ for all $\omega \in W_p(q)$ since in this case d^i is a quasi-subgradient of h at ω^i. On the other hand, if $p(\omega^i) \notin R^k_+$ then $W_p(q) \subset \{\omega : p(\omega) \geq P\_(p(\omega^i))\}$ because $R^k_+ \subset P\_(p(\omega^i)) + R^k_+$. Then (IP) implies that $\langle d^i, \omega \rangle \geq \langle d^i, \omega^i \rangle$ for all $\omega \in W_p(q)$ since in that case d^i is a supporting element of p at ω^i. Thus in both cases the inequality $\langle d^i, \omega \rangle \geq \langle d^i, \omega^i \rangle$ holds for all $\omega \in W_p(q)$.

From this moment on we can follow exactly the proof of the main theorem in [24]. Considering the point $\omega = \tilde{\omega} - \delta d^i / \|d^i\| \in W_p(q)$, for which the above inequality holds, we obtain the estimate

$\|\omega^{i+1}-\tilde{\omega}\|^2 \leqslant \|\omega^i-\tilde{\omega}\|^2 + t_i^2 - 2t_i\delta$. Choosing sufficiently large I to assure $t_i < \delta$ for all $i \geqslant I$ and summing up m consecutive estimates of the above type, starting from I, we derive: $0 < \|\omega^I-\tilde{\omega}\|^2 - \delta \sum_{i=I}^{I+m} t_i$. This inequality, holding for any m, contradicts the assumption that the series Σt_i is divergent.

Thus the supposition leads to the contradiction, hence there exists $\omega^i \in W_p(q)$ for any $q < \hat{h}$. If we choose the sequence $q_s \to \hat{h}$ for $s \to \infty$ then the corresponding sequence $\omega^{i_s} \in W_p(q_s)$ is such that $h(\omega^{i_s}) \to \hat{h}$ for $s \to \infty$.

□

The above theorem states the main result concerning the convergence of algorithms, designed to solve (MP) and based on the concept of quasi-subgradient, which are presented in the continuation of this chapter.

The surrogate dual problem (SD) can be directly represented by (MP) if we substitute D* for G and if we neglect p. All assumptions in (MP) are satisfied then and h is the surrogate dual function. We need a quasi-subgradient of h at each point $\omega \in D^*$ to apply the recursive scheme (RS). There is no need for any additional direction since the constraints represented by p do not appear.

Proposition 3.2

If $\bar{x} \in Q(\omega)$ (this denotes that $\bar{x}$ is an optimal solution for (S)) then $g(\bar{x})$ is a quasi-subgradient of h at $\omega \in D^*$.

Proof

$h(\omega)=f(\bar{x})$ and $\langle g(\bar{x}),\omega\rangle \leqslant 0$ from the statement of (S). Consider $u \in D^*$ such that $\langle g(\bar{x}),u\rangle \leqslant \langle g(\bar{x}),\omega\rangle$. This means that $\bar{x} \in X(u)$. Hence $h(u)=\min\{f(x) : x \in X(u)\} \leqslant f(\bar{x})=h(\omega)$. Thus, the implication (IQ) is shown for $g(\bar{x})$.

□

The above result guarantees the existence of a quasi-subgradient at each point in D* and moreover shows how to construct it by solving the surrogate problem. The scheme (RS) can be used for the surrogate dual problem. It takes on the following form:

$$\omega^{i+1} = P_{D^*}(\omega^i + t_i g^i/\|g^i\|),\quad \omega^1 \in D^*$$

where $g^i=g(x^i)$ for $x^i \in Q(\omega^i)$.

Proposition 3.1 implies that the value $\hat{h}$ can be approximated with an arbitrary accuracy if the above scheme is applied. The surrogate problems must be solved for all $\omega^i \in D^*$ to calculate $h(\omega^i)$ and g^i.

Choosing the sequence $\{\omega^i\}$ from an unbounded set is the main disadvantage of the above approach from the operative point of view. This may cause poor behaviour of any algorithm based on it in the process of numerical calculations. The properties of the function h offer the opportunity of restricting the feasible set in (SD) to a compact set. Accordingly the surrogate dual problem will be reformulated in the next section.

4. Three algorithms

At first we consider the case where only inequality constraints are present in the primal problem (P). Then $D=R^m_+$ and $D^*=R^m_+$. An arbitrary normalization of the surrogate dual multipliers is possible and leads to a bounded feasible set of them. Hence, we assume that the maximization of the surrogate dual function proceeds on the set

$$K_+ = \{\omega \in R^m_+ : \ ||\omega|| = 1\}$$

This does not restrict the range of h except that the inessential value $h(0)$ is excluded. The value of $h(\omega)$ is constant along the line $\{u : u=\alpha\omega \ \text{ for } \ \alpha > 0\}$ for any $\omega \in R^m_+$. This means that the modification of quasi-subgradient proposed in [4] better corresponds to that property of h since it provides a quasi-subgradient orthogonal to this line.

<u>Proposition 4.1</u>

<u>If</u> $\bar{x} \in Q(\omega)$ <u>then</u> $d(\bar{x}) = g(\bar{x}) - \langle g(\bar{x}), \omega \rangle \omega$ <u>is a quasi-subgradient of</u> h <u>at</u> $\omega \in R^m_+$ <u>such that</u> $||\omega|| = 1$.

<u>Proof</u>

Assume at first that $d(\bar{x}) \neq 0$. Consider $u \in R^m_+$ such that $\langle d(\bar{x}), u \rangle \leqslant \langle d(\bar{x}), \omega \rangle$. The right hand side of the inequality is equal to zero so $\langle g(\bar{x}), u \rangle \leqslant \langle g(\bar{x}), \omega \rangle \ \langle \omega, u \rangle$. Proposition 3.2 and the implication (IQ) imply that the inequality $h(u) \leqslant h\ (\langle \omega, u \rangle \omega)$ holds. The property of h (see Section 2) indicates that $h(\langle \omega, u \rangle \omega) \leqslant h(\omega)$. This shows that (IQ) holds for $d(\bar{x}) \neq 0$.

Consider now the case $d(\bar{x}) = 0$. Then $g(\bar{x}) = \langle g(\bar{x}), \omega \rangle \omega$. The statement of (S) indicates that $\langle g(\bar{x}), \omega \rangle \leqslant 0$ so $g(\bar{x}) \leqslant 0$ $(\omega \in R^m_+)$. We find that $d(\bar{x}) = 0$ implies the feasibility of $\bar{x}$ in the primal problem (P). Then $h(\omega) = \hat{h} = \hat{f}$ i.e. ω is an optimal surrogate multiplier and there is no duality gap. This means $d(\bar{x})=0$ is a quasi-subgradient of h at ω.

□

Thus we can use at each point $\omega \in K_+$ a quasi-subgradient which is orthogonal to the direction of ω ($\langle d(x),\omega\rangle = 0$ for $x \in Q(\omega)$) and hence we can construct a search direction which is tangent to the sphere $S(0;1)$ at ω.

The surrogate dual problem in the version discussed here takes on the form

$$\hat{h} = \sup\{h(\omega) : \omega \in K_+\}$$

In order to apply the scheme (RS) to this program we represent it in terms of (MP). We substitute the unit ball $B(0;1)$ for G and we set $p(\omega)=\omega$, $k=m$.

Suppose that in some iteration of (RS) a point ω^i has been generated and is feasible subject to p (i.e. $\omega^i \in R^m_+$). According to the scheme the step t_i is made along the direction $\bar{d}^i = d^i/\|d^i\|$, where d^i is a quasi-subgradient of h at ω^i (based on Proposition 4.1 we use $d^i=d(x^i)$ for $x^i \in Q(\omega^i)$ if additionally $\|\omega^i\| = 1$). If the point $\bar{\omega}^i = \omega^i + t_i\bar{d}^i$ is infeasible i.e. $\bar{\omega}^i \notin R^m_+$, then in the next iteration of the scheme we use a supporting element of the identity function p with respect to R^m_+ at $\bar{\omega}^i$. The set of indices $J^i=\{j : \bar{\omega}^i_j < 0\}$ is non-empty because $\bar{\omega}^i \notin R^m_+$. The subgradient of the j-th component of p has the form $\bar{p}_j=[0\ldots1\ldots0]^T$, where the j-th coordinate is equal to one. We define the positive parameters $\alpha^i_j = -\bar{\omega}^i_j$ for $j \in J^i$. Hence we can apply Lemma 3.2 to construct $\tilde{d}^i = \sum_{j\in J^i} \alpha^i_j \bar{p}_j$ which is a supporting element of p at $\bar{\omega}^i$.

$$\tilde{d}^i_j = \begin{cases} \alpha^i_j, & \text{if } j \in J^i \\ 0, & \text{else} \end{cases} \qquad \text{for } j=1,\ldots,m$$

Thus we can make the next iteration. If we set the step size $\tilde{t}_i = 2\sqrt{\sum_{j\in J^i} (\bar{\omega}^i_j)^2}$ then the point $\tilde{\omega}^i = \bar{\omega}^i + \tilde{t}_i\tilde{d}^i/\|\tilde{d}^i\|$ is feasible subject to p bacause $\tilde{\omega}^i_j = |\bar{\omega}^i_j| > 0$ for $j=1,\ldots,m$. Moreover $\tilde{t}_i \leq 2t_i\sqrt{m}$ so it tends to zero if $i \to \infty$.

The direction d^i is tangent to $S(0;1)$ at ω^i and hence $\|\tilde{\omega}\| > 1$ if only $\|\omega^i\| = 1$. Thus the projection on $G=B(0;1)$ can be calculated by normalization of $\tilde{\omega}^i$. The next point $\omega^{i+1}=\tilde{\omega}^i/\|\tilde{\omega}^i\|$ belongs to K_+. If we choose the starting point $\omega^1 \in K_+$ then all points generated by the recursive scheme (RS) remain in $K_+ : \{\omega^i\} \subset K_+$.

The above construction can be formulated as an implementable algorithm which follows the scheme (RS). Additionally this algorithm is endowed with an effective stop test which is based on Corollary 2.2.

Algorithm A

Step 0: Set $\gamma > 0$, $\omega^1 \in K_+$, $Z_0 \leftarrow \emptyset$, $q_1 \leftarrow -\infty$. Choose a positive sequence $\{t_i\}_1^\infty$ such that $t_i \to 0$ and $\sum_{i=1}^{\infty} t_i = \infty$.

Step 1: Determine any $x^i \in Q(\omega^i)$. Let $g^i = g(x^i)$.

Step 2: If $f(x^i) > q_i$ then $q_i \leftarrow f(x^i)$.

Step 3: $Z_i \leftarrow Z_{i-1} \cup \{g^i\}$, $Y_i \leftarrow \mathrm{co} Z_i$.

Step 4: If $Y_i \cap -R_+^m \neq \emptyset$ then stop.

Step 5: $d^i \leftarrow g^i - \langle g^i, \omega^i \rangle \omega^i$ and $\bar{d}^i \leftarrow d^i / \|d^i\|$.

Step 6: $\bar{\omega}^i \leftarrow \omega^i + \gamma t_i \bar{d}^i$ and $\tilde{\omega}_j^i \leftarrow |\bar{\omega}_j^i|$ for $j=1,\dots,m$.

Step 7: $\omega^{i+1} \leftarrow \tilde{\omega}^i / \|\tilde{\omega}^i\|$, $i \leftarrow i+1$. Go to Step 1.

Algorithm A strictly corresponds to the algorithm presented in [4] except that the new stop test is added.

Proposion 3.1 implies

Corollary 4.1

$\lim_{i\to\infty} q_i = \hat{h}$, where the sequence $\{q_i\}$ is generated in Algorithm A.

The stop test in Algorithm A is designed to recognize the situation in which the value $\hat{h}$ is attained. In [4] only the condition $g^i \in -R_+^m$ is checked to assure that $d^i \neq 0$. This allows us to check whether $q_i = \hat{h}$ only if $\hat{h} = \hat{f}$. In many practical problems we cannot expect that there is no surrogate duality gap so such a test can be active rather rarely. Let us recall the second characterization of $\hat{h}$ (Corollary 2.2) which takes now on the form

$$\hat{h} = \min\{q : \mathrm{co}R(q) \cap -R_+^m \neq \emptyset\}, \text{ where}$$

$$R(q) = \{g(x) : x \in X_0,\ f(x) \leq q\}$$

In general, we are not able to describe numerically the whole the set $R(q_i)$ in each iteration of Algorithm A. However, notice that g^i generated in each iteration belongs to $R(q_i)$ since $q_i = \max\{f(x^s) : s=1,\dots,i\}$. Thus the set $Z_i = \{g^1,\dots,g^i\}$ can be used to generate an inner approximation of $\mathrm{co}R(q_i)$. Hence $Y_i \subset \mathrm{co}R(q_i)$ and if $Y_i \cap -R_+^m \neq \emptyset$ then obviously $q_i = \hat{h}$. Also if any $g^i \in -R_+^m$ then clearly the intersection is nonempty. This condition can be checked rather easily using a linear program to calculate a distance between Y_i and $-R^m_+$; denote it by r_i. The condition $r_i=0$ is equivalent to $Y_i \cap -R^m_+ \neq \emptyset$. In practice, the stop test can be used in the form: $r_i < \varepsilon$, where ε is a given accuracy. This is validated by the result reported in [26]: $\lim_{i\to\infty} r_i=0$ for Algorithm A.

We can modify some elements in the presentation which preceedes the formulation of Algorithm A. This leads to a slightly different algorithm which shows that not only one algorithm can be interpreted in the sense of the recursive scheme (RS). At first, notice that the element $\bar{g}(\bar{x})=g(\bar{x})+b(\bar{x})$ for $\bar{x} \in Q(\omega)$ is also a quasi-subgradient of h at ω if

$$b_j(\bar{x}) = \begin{cases} -g_j(\bar{x}), & \text{if } g_j(\bar{x}) < 0 \text{ and } \omega_j=0 \\ 0, & \text{otherwise} \end{cases}$$

This follows from the proof of Proposition 3.2 since by adding $b(\bar{x})$ we do not change the value of the scalar product $\langle g(\bar{x}),\omega\rangle = \langle \bar{g}(\bar{x}),\omega\rangle$. Thus instead of g^i we can use $\bar{g}^i = \bar{g}(x^i)=g(x^i)+b(x^i)$ to calculate d^i i.e. $d^i = \bar{g}^i - \langle\bar{g}^i,\omega^i\rangle\omega^i$.

Moreover the step size t_i can be halved and the point $\tilde{\omega}^i = \bar{\omega}^i + 0.5\,\tilde{t}_i\,\tilde{d}^i/\|\tilde{d}^i\|$ remains feasible subject to p since $\tilde{\omega}^i_j=\max(0,\bar{\omega}^i_j) > 0$ for $j=1,\dots,m$. The algorithm obtained as a result of such modifications takes on the following form:

<u>Algorithm B</u>

Step 0-4: as in Algorithm A.

Step 5: Determine $\bar{g}^i = g^i + b^i$, where $b^i_j = \begin{cases} -g^i_j, & \text{if } g^i_j<0 \text{ and } \omega^i_j=0 \\ 0, & \text{otherwise} \end{cases}$

Step 6: $d^i \leftarrow \bar{g}^i - \langle\bar{g}^i,\omega^i\rangle\omega^i$ and $\bar{d}^i \leftarrow d^i/\|d^i\|$.

Step 7: $\bar{\omega}^i \leftarrow \omega^i + t_i\bar{d}^i$, $\tilde{\omega}^i_j \leftarrow \max(0,\bar{\omega}^i_j)$ for $j=1,\dots,m$.

Step 8: $\omega^{i+1} \leftarrow \tilde{\omega}^i/\|\tilde{\omega}^i\|$, $i \leftarrow i+1$. Go to Step 1.

When $\omega^i \in \text{int } R^m_+$ and t_i is so small that $\bar{\omega}^i \in R^m_+$ both algorithms yield the same point as a new approximation. The difference appears when a face of the positive orthant becomes "active" i.e. $\bar{\omega}^i_s < 0$ for some s. While Algorithm A jumps back from such a constraint: $\omega^{i+1}_s>0$, Algorithm B puts the new approximation exactly on this hyperplane: $\omega^{i+1}_s=0$. The latter allows also searching along such an "active" hyperplane due to Step 5. This difference is illustrated in Fig.3 which shows a situation of two negative coordinates of $\bar{\omega}^i$. Nevertheless both algorithms are based on the same scheme (RS) and Corollary 4.1 is valid for Algorithm B too. Also all remarks about the stop test remain true.

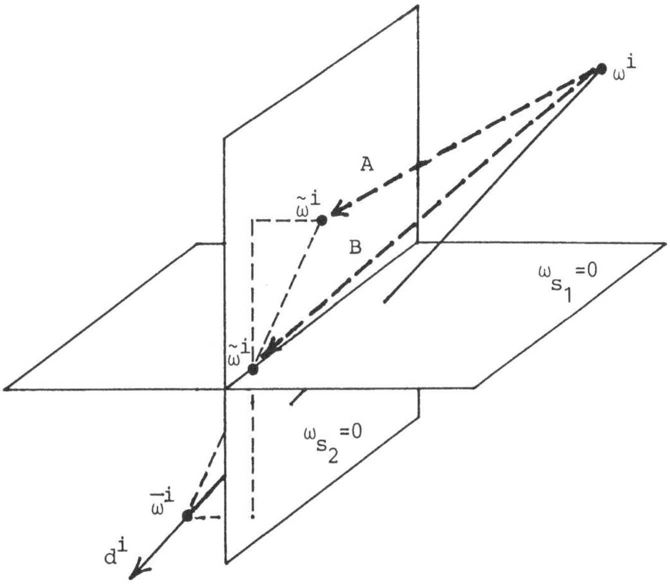

Fig. 3

Consider now the case where in the primal problem (P) only equality constraints are present. Then $D=\{0\}$ and $D^*=R^m$. Thus initially there is no restriction on dual multipliers and (SD) becomes an unconstrained maximization problem. An arbitrary normalization of them is obviously possible and helps to establish a bounded feasible set for the surrogate dual problem. On that purpose we propose the set

$$K = \{\omega \in R^m : \ ||\omega|| \leqslant 1\}$$

We can calculate a quasi-subgradient of h at any point ω from K on the basis of Proposition 3.2. Proposition 4.1 cannot be applied because the element $d(\bar{x})$ for $\bar{x} \in Q(\omega)$ is not necessarily a quasi-subgradient of h. The inclusion $\{u : \langle d(\bar{x}), u\rangle \leqslant \langle d(x), \omega\rangle\} \subset \{u : \langle g(\bar{x}), u\rangle \leqslant \langle g(x), \omega\rangle\}$ does not hold except in the case $\langle g(\bar{x}), \omega\rangle = 0$. Hence, $g(\bar{x})$ must be used directly. This requires to include in K also multipliers with norm less than one.

Thus we should like to apply the scheme (RS) to the problem

$$h = \sup\{h(\omega) : \omega \in K\}$$

To realize this purpose it is enough to substitute the unit ball $B(0;1)$ for G, as in the previous case, and to neglect constraints given by p in (MP). Then we can follow (RS) directly and never use the second possibility to choose d^i as a supporting element. The projection on $B(0;1)$ required in the recursive scheme is calculated by the normalization if a current multiplier lies outside of K otherwise it is just the identity transformation.

If $g(x) = 0 \in R^m$ for $x \in Q(\omega^i)$ then the course of generation of $\{\omega^i\}$ can be interrupted because this means $h(\omega^i)=\hat{h}=\hat{f}$. For a more general stop test Corollary 2.2 can be utilized. It takes now on the form:

$$h = \min\{q : 0 \in \mathrm{co}R(q)\}, \text{ where}$$

$$R(q) = \{g(x) : x \in X_o,\ f(x) \leqslant q\}$$

It is still true that the set $R(q)$ cannot be described numerically in each iteration so an inner approximation of $\mathrm{co}R(q)$ is proposed. The following algorithm represents (RS) with the additional stop test.

Algorithm C

Step 0: Set $\gamma > 0$, $\omega^1 \in K$, $Z_o \leftarrow \emptyset$, $q_1 \leftarrow -\infty$. Choose a positive sequence $\{t_i\}_1^\infty$ such that $t_i \to 0$ and $\sum_{i=1}^{\infty} t_i = \infty$.

Step 2: If $f(x^i) > q_i$ then $q_i \leftarrow f(x^i)$.

Step 3: $Z_i \leftarrow Z_i \cup \{g^i\}$, $Y_i \leftarrow \mathrm{co}Z_i$.

Step 4: If $0 \in Y_i$ then stop.

Step 5: $d^i \leftarrow g^i / \|g^i\|$, $\bar{\omega}^i \leftarrow \omega^i + \gamma t_i d^i$.

Step 6: $\omega^{i+1} = \begin{cases} \bar{\omega}^i, & \text{if } \|\omega^i\| \leqslant 1 \\ \bar{\omega}^i / \|\bar{\omega}^i\|, & \text{otherwise} \end{cases}$, $i \leftarrow i+1$ and go to Step 1.

The main convergence theorem for (RS) implies

Corollary 4.2

$\lim_{i\to\infty} q_i = \hat{h}$, where $\{q_i\}$ is generated in Algorithm C.

Each quasi-subgradient g^i belongs to $R(q_i)$ by the same reason as in Algorithm A and B. Hence $Y_i \subset \mathrm{co}R(q_i)$. Thus, if $0 \in Y_i$ then $q_i = \hat{h}$. The stop test is obviously active also if $g^i=0$. The condition $0 \in Y_i$ can be checked with a linear program which yields a distance between 0 and Y_i. Denote this distance by r_i. It is necessary to assure that $r_i \to 0$ for $i \to \infty$ if we intend to use in practice the stop test $r_i < \varepsilon$.

Lemma 4.1

If $u \in K$ then $\liminf_{i\to\infty} \langle g^i, u \rangle \leqslant 0$, where g^i is calculated in the i-th iteration of Algorithm C.

Proof

$\|\bar{\omega}^i - u\|^2 \leqslant \|\omega^i - u\|^2 + t_i'(t_i' - 2\langle d^i, u\rangle)$ since $\langle d^i, \omega^i\rangle = \langle g^i, \omega^i\rangle / \|g^i\| \leqslant 0$ $(t_i' = \gamma t_i)$. In the case where $\|\bar{\omega}^i\| \leqslant 1$ we set $\omega^{i+1} = \bar{\omega}^i$. Consider the case where $\|\bar{\omega}^i\| > 1$. Then $\|\omega^{i+1}\| = 1$ and $\langle u, \omega^{i+1}\rangle \leqslant 1$ for all $u \in K$. This implies $\|\bar{\omega}^i - u\|^2 \geqslant \|\omega^{i+1} - u\|^2 + (\|\bar{\omega}^i\| - 1)^2 \geqslant \|\omega^{i+1} - u\|^2$. Thus in both cases $\|\omega^{i+1} - u\|^2 \leqslant \|\bar{\omega}^i - u\|^2 \leqslant \|\omega^i - u\|^2 + t_i'(t_i' - 2\langle d^i, u\rangle)$. Moreover the sequence $\{g^i\}$ is bounded and so is the sequence $\{\langle d^i, u\rangle\}$.

Suppose on the contrary that there exists $\delta > 0$ such that $\langle d^i, u\rangle \geqslant \delta$ for all $i \geqslant I_1$. Choose sufficiently large I_2 to assure that $t_i' \leqslant \delta$ for $i \geqslant I_2$. Then $\|\omega^i - u\|^2 - \|\omega^{i+1} - u\|^2 \geqslant t_i'(2\langle d^i, u\rangle - t_i') \geqslant t_i'\delta$ for $i \geqslant I = \max(I_1, I_2)$. If we sum up k inequalities of this type, starting from I, then we obtain that $2 \geqslant \|\omega^I - u\|^2 - \|\omega^{I+k} - u\|^2 \geqslant \sum_{i=I}^{i+k} t_i'$. This condition is to be true for any k so it contradicts the assumption that the series Σt_i is divergent. This completes the proof.

□

On the basis of the above result the required property of $\{r_i\}$ can be proved.

Proposition 4.2

The sequence $\{r_i\}_1^\infty$ generated in Algorithm C by calculation of the distance between 0 and Y_i contains an element $r_{i_o} = 0$ or satisfies the condition $\lim_{i\to\infty} r_i = 0$.

Proof proceeds analogically to the proof of the corresponding theorem for Algorithm A (or B) [26] with the only difference that the point 0 and the set Y_i are separated strictly by a hyperplane defined by $u \in K \setminus \{0\}$.

□

The above theorem not only validates the stop test with a given accurary $r_i < \varepsilon$, but also suggests that the distance r_i can indicate indirectly how well $\hat{h}$ is approximated by q_i. This remark refers as well to both algorithms constructed in the case $D=R^m_+$.

All algorithms presented in this section of the chapter have their forms suitable for implementation. Obviously some parts of them can take different form while being still based on the same recursive scheme (RS). Also the surrogate dual problem can be represented by (MP) in a few different ways. Thus, it is not claimed here that Algorithms A, B and C are unique or even best representations of the scheme (RS). They are described as examples in order to show how the quasi-subgradient method of maximization can be applied to the surrogate dual problem.

5. Numerical experiments

Two integer programming test problems with linear inequality constraints were solved using Algorithms A and B. They have the same form:

$$\hat{f} = \min\{ \sum_{k=1}^{n} f_k(x_k) : x \in X\}, \quad X = \{ x \in X_o : \sum_{k=1}^{n} a_{jk}x_k \leqslant b_j, \; j=1,\dots,m\}.$$

$$X_o = \{x \in Z^n, \; 0 \leqslant x_k \leqslant \psi_k, \qquad k=1,\dots,n\}$$

Two similar test problems were solved using Algorithm C. They have objective function of the same separable form and also linear constraints but of equality type, namely:

$$X = \{x \in X_o : \sum_{k=1}^{n} a_{jk}x_k = b_j, \quad j=1,\dots,m\}$$

Each objective function is represented by a table $F=[f_{ks}]$: $f_{ks} = f_k(s)$, $s=1,\dots,\psi_k$ $(f_k(0)=0)$. Elements of these tables are real numbers randomly chosen from the interval $[-9.9; 0.0]$. Integer elements of matrix $A=[a_{jk}]$ were randomly chosen from the set $\{0,1,\dots,9\}$. Vector b was determined for each test problem in such a way that the unrestricted minima of the objective function were infeasible. All numerical data are included in the Appendix.

The sequence $\{1/i\}$, where i is an iteration number, was used as the required step size sequence for all tested algorithms. Each test problem was solved using each prescribed algorithm for five different values of the parameter γ: 0.25, 0.5, 1.0, 2.0 and 4.0. The value $\varepsilon = 10^{-5}$ was taken as a stop test accuracy for all runs during the experiments.

The starting point $\omega_j^1 = 1/\sqrt{m}$, $j=1,\dots,m$, was set for Algorithms A and B. In the case of equality constraints $\omega_j^1 = 0$, $j=1,\dots,m$, were chosen. These starting points are neutral and they do not prefer a priori any primal constraints. The first one includes each primal constraint in the starting surrogate constraint with the same coefficient while the latter neglects all of them and the starting surrogate problem is unconstrained in that sense. If some information about "activity" of primal inequality constraints is available before initiation of the algorithm then some of them can be neglected by setting $\omega_j^1=0$ for all coordinates corresponding to the "non-binding" ones. This can be especially valuable for Algorithm B because it leads to searching in a subset of the dual feasible set K having smaller dimension. If some information of the same kind can be obtained about primal equality constraints then the starting point in Algorithm C can be chosen in the proper orthant and it will be closer to the dual optimal set.

<u>Test 1</u> (n=20, m=15)

Figure 1.1 shows the values $h(\omega^i)$ generated in the best runs

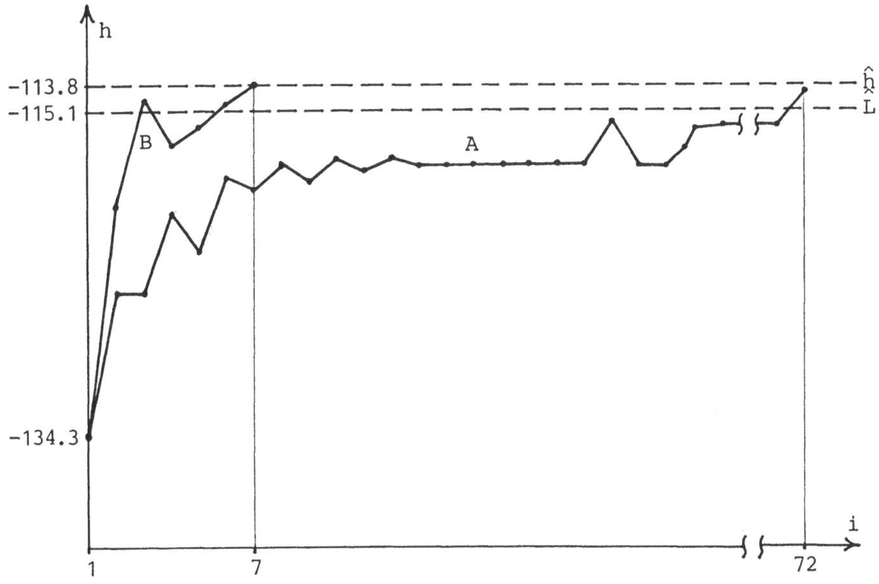

Fig. 1.1

of Algorithms A and B. Algorithm A was terminated for $\gamma=2.0$ in 75th iteration because its stop test was still inactive: $r_{75}=1.3112$ ($r_1=96.0$). The best lower bound on the unknown $\hat{f}$ was $q_{75}=h(\omega^{72})=-113.90$. Algorithm B was terminated for $\gamma=2.0$ in 10th iteration by its stop test: $r_{10}<10^{-5}$. Most probably $\hat{h}=q_{10}=h(\omega^7)=-113.80$. The value $\hat{L}\approx -115.1483$ calculated numerically is also marked in the figure.

j=	ω		v
	A	B	
1	0.0013	0.0	0.0
2	0.0061	0.0	0.0
3	0.0024	0.0	0.0
4	0.9361	0.9484	0.5305
5	0.0009	0.0	0.0
6	0.0367	0.0	0.0
7	0.3494	0.3172	0.2120
8	0.0051	0.0	0.0
9	0.0004	0.0	0.0
10	0.0012	0.0	0.0
11	0.0105	0.0	0.0
12	0.0020	0.0	0.0
13	0.0064	0.0	0.0
14	0.0020	0.0	0.0
15	0.0024	0.0	0.0

$h(v) = -113.90$

Table 1.1

i=	x		
	A	B	Lagr.
1	0	0	0
2	3	3	0
3	1	1	1
4	1	1	1
5	1	1	1
6	2	2	2
7	1	0	1
8	1	1	1
9	1	0	1
10	1	1	1
11	1	1	1
12	1	1	1
13	1	1	1
14	1	1	1
15	1	1	1
16	0	2	2
17	1	1	1
18	0	1	1
19	0	0	0
20	0	0	0

Table 1.2

Table 1.1 contains the vectors: ω^{72} (for Alg.A), ω^{7} (for Alg.B) which correspond to the maximal values of $h(\omega^i)$ and v which corresponds to $\hat{L}$. Table 1.2 presents optimal solutions found in the surrogate problems (S) which gave the best bounds and an optimal solution to the Lagrangean problem (L) with the optimal value $\hat{L}$.

<u>Test 2</u> (n=20, m=20)

Figure 2.1 presents the sequences $\{h(\omega^i)\}$ from the best runs.

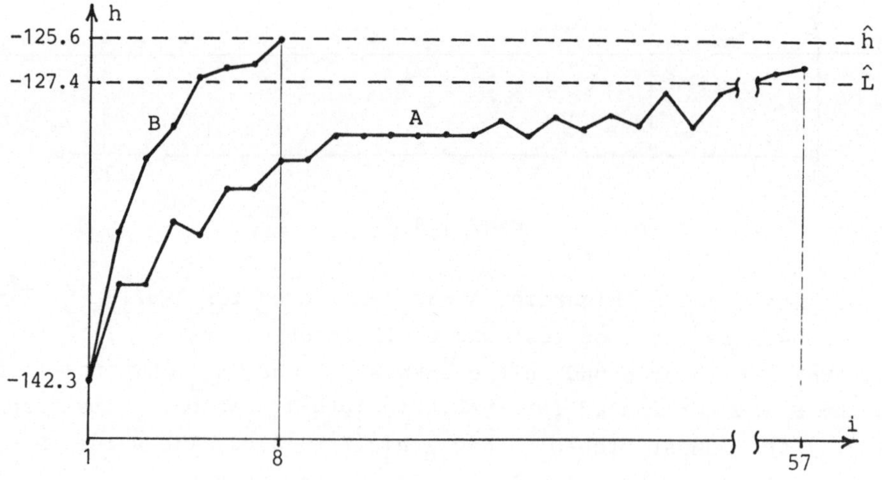

Figure 2.1

j=	ω		v
	A	B	
1	0.0045	0.0289	0.0
2	0.0004	0.0	0.0
3	0.0135	0.0	0.0
4	0.0029	0.0	0.0
5	0.5341	0.5789	0.1524
6	0.0	0.0	0.0
7	0.0023	0.0	0.0
8	0.0005	0.0	0.0
9	0.0040	0.0	0.0
10	0.0044	0.0	0.0
11	0.0010	0.0	0.0
12	0.0039	0.0	0.0
13	0.0017	0.0	0.0
14	0.0028	0.0078	0.0
15	0.0059	0.0	0.0
16	0.0023	0.0	0.0
17	0.0049	0.0	0.0
18	0.2247	0.1399	0.0817
19	0.0019	0.0	0.0
20	0.8148	0.8027	0.2518

$h(v) = -125.60$

Table 2.1

i=	x		
	A	B	Lagr.
1	3	3	2
2	0	1	0
3	1	1	1
4	1	1	1
5	1	1	1
6	0	0	2
7	1	1	1
8	1	1	1
9	3	1	1
10	1	1	1
11	1	1	1
12	1	1	1
13	1	1	1
14	1	1	1
15	1	1	1
16	0	0	0
17	1	1	1
18	4	4	4
19	2	2	2
20	0	0	0

Table 2.2

The value $\hat{L} \approx -127.4359$ is also marked. Algorithm A was terminated for $\gamma=2.0$ in 75th iteration and its stop test was inactive: $r_{75}=3.0002$ ($r_1=172.0$). The best lower bound was $h(\omega^{57})=-126.60$. Algorithm B was terminated for $\gamma=1.0$ in 13th iteration by the test: $r_{13} < 10^{-5}$. Hence $\hat{h}=q_{13}=h(\omega^{8})=-125.60$. Tables 2.1 and 2.2 are of the same form as Tables 1.1 and 1.2.

<u>Test 3</u> (n=5, m=3)

Figure 3.1 presents the sequence $\{h(\omega^i)\}$ from the best run.

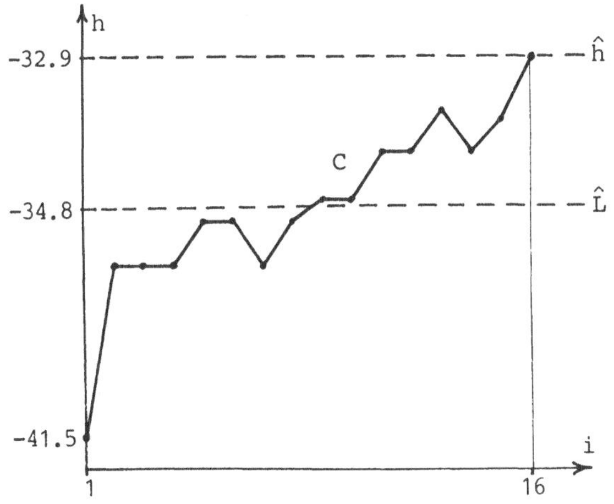

Figure 3.1

j=	ω	v
1	0.3970	0.5026
2	-0.5815	-0.9269
3	0.4064	0.7362

Table 3.1

i=	x	
	C	Lagr.
1	1	1
2	1	1
3	4	1
4	5	5
5	0	1

Table 3.2

Algorithm C was terminated for $\gamma=1.0$ by its stop test: $r_{22} < 10^{-5}$. Hence $\hat{h}=q_{22}=h(\omega^{16})=-32.90$. The value $\hat{L} \approx -34.8126$ is marked in the figure.

Tables 3.1 and 3.2 are analogous to Tables 1.1 and 1.2, respectively.

<u>Test 4</u> (n=10, m=5)

The presentation of results is the same as in Test 3: Figure 4.1, Tables 4.1 and 4.2 give details of the best run. Algorithm C was terminated for $\gamma=2.0$ by its stop test: $r_{25} < 10^{-5}$. Hence $\hat{h}=q_{25}=h(\omega^{14}) = -78.50$. The value $\hat{L} \approx -79.1267$.

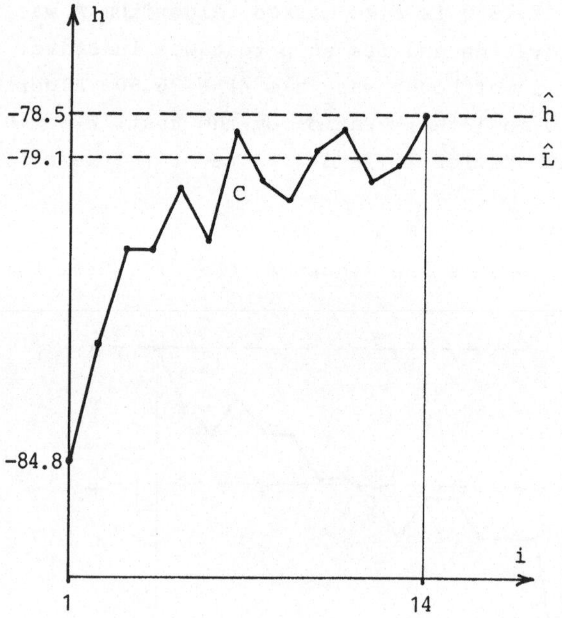

Figure 4.1

j=	ω	v
1	0.2924	0.0523
2	0.4973	0.0921
3	0.3189	0.0603
4	-0.0897	-0.0221
5	-0.7466	-0.1698

Table 4.1

i=	x	
	C	Lagr.
1	1	1
2	1	1
3	4	4
4	5	5
5	4	4
6	1	1
7	4	4
8	1	1
9	1	1
10	2	5

Table 4.2

The numerical experiments were performed at the Computer Center of the University of Hiroshima during the author stay as a visiting research associate at the University on leave from the Polish Academy of Sciences in the period between October 1982 and March 1984.

The author would like to express his gratitude to Professor K. Mizukami for his benevolence and help.

6. Concluding remarks

The surrogate dual problem described in Section 2 yields usually a smaller duality gap than the Lagrangean dual problem in the case of nonconvex problems. Hence the approach presented in this paper may be more efficient in the process of bounding in branch-and-bound methods of solving integer programming problems. However, it leads to the solution of a more difficult mathematical program with a quasiconcave objective function.

The theoretical results presented in Section 2 and 3 form a basis for algorithms of calculating surrogate constraints. The numerical experiments indicate that some of them may be useful in practice for bounding the primal optimal value, for example Algorithm B. Algorithm C shows that even in the case of equality-constrained primal problems such a bound can be calculated iteratively using surrogate inequalities.

References

[1] P.L. Bricker: Bounding a class of nonconvex linearly-constrained resource allocation problems via the surrogate dual, Mathematical Programming 18 (1980) 68-83.

[2] J.-P. Crouzeix: A duality framework in quasiconvex programming, in Generalized Concavity in Optimization and Economics, eds. Shaible S., Ziemba W.T., Academic Press, New York, (1981), 207-225.

[3] J.J. Dinkel, G.A. Kochenberger: An implementation of surrogate

constraint duality, Operations Research 26 (1978) 358-364.

[4] M.E. Dyer: Calculating surrogate constraints, Mathematical Programming 19 (1980) 255-278.

[5] W. Findeisen, J. Szymanowski, A. Wierzbicki: Theory and Computational Methods of Optimization, PWN, Warszawa, 1977, (in Polish).

[6] M.L. Fisher, W.D. Northup, J.F. Shapiro: Using duality to solve discrete optimization problems: theory and computational experience, Mathematical Programming Study 3 (1975) 56-94.

[7] M.L. Fisher, B.J. Lageweg, J.K. Lenstra, A.H.G. Rinooy kan: Surrogate duality relaxation for job shop scheduling, Discrete Applied Mathematics 5 (1983) 1, 65-75.

[8] R.S. Garfinkel, G.L. Nemhauser, Integer Programming, John Wiley & Sons, New York, 1972.

[9] A.M. Geoffrion: Lagrangean relaxation and its uses in integer programming, Mathematical Programming Study 2 (1974) 82-114.

[10] F. Glover: A multiphase-dual algorithm for the zero-one-integer programming problem, Operations Research 13 (1965) 879.919.

[11] F. Glover: Surrogate constraints, Operations Research 16 (1968), 741-749.

[12] F. Glover: Surrogate constraint duality in mathematical programming, Operations Research 23 (1975) 434-451.

[13] H.J. Greenberg, W.P. Pierskalla: Surrogate mathematical programs, Operations Research 18 (1970) 924-939.

[14] H.J. Greenberg, W.P. Pierskalla: A review of quasi-convex functions, Operations Research 19 (1971) 1553-1570.

[15] H.J. Greenberg, W.P. Pierskalla: Quasi-conjugate functions and surrogate duality, Cahiers du Centre d'Etudes de Recherche Operationnelle 15 (1973) 437-448.

[16] H.J. Greenberg: The generalised penalty function surrogate model, Operations Research 21 (1973) 162-178.

[17] H. Held, P. Wolfe, H.P. Crowder: A validations of subgradient optimisation, Mathematical Programming 6 (1974) 62-88.

[18] M.H. Karwan, R.L. Rardin: Some relationships between Lagrangean and surrogate duality in integer linear programming, Mathematical Programming 17 (1979) 320-334.

[19] M.H. Karwan, R.L. Rardin: Surrogate duality in a branch-and-bound procedure, Naval Research Logistics Quarterly 28 (1981) 1, 93-101.

[20] M.H. Karwan, R.L. Rardin: Surrogate dual multiplier search procedures in integer programming, Operations Research 32 (1984), 52-69.

[21] M.H. Karwan, R.L. Rardin, S. Sarin: A new surrogate dual multiplier search procedure, Technical Report, August 1984, North Carolina A. a. T. State University, Grensboro.

[22] D.G. Luenberger: Quasi-convex programming, SIAM Journal of Applied Mathematics 16 (1968) 1090-1095.

[23] W. Oettli: Optimality conditions involving generalized convex mappings, in Generalized Concavity in Optimization and Economics, eds. Schaible S., Ziemba W.T., Academic Press, New York, (1981), 227-238.

[24] B.T. Polyak: A general method of solving extremal problems,

Doklady Akademi Nauk SSSR 174 (1967) 33-36.

[25] R.T. Rockafellar: Convex Analysis, Princeton University Press, Princeton, New Jersey, 1970.

[26] J. Sikorski, K. Mizukami: Three algorithms for calculating surrogate constraints in integer programming problems, Control & Cybernetics 13 (1984) 375-397.

Appendix

Test 1 $\psi_k = 5$ for $k=1,\dots,20$

F:

f_{ks}	1	2	3	4	5
1	0.0	-6.3	-8.9	-5.7	-7.3
2	-1.1	-2.8	-7.5	-5.6	-1.2
3	-7.5	-2.7	-4.1	-6.4	-7.5
4	-8.0	-1.7	-8.9	-5.8	-1.1
5	-9.1	-7.3	-2.3	-8.5	-6.6
6	-0.1	-8.1	-2.7	-0.8	-4.8
7	-5.2	-0.1	-4.6	-4.3	-8.8
8	-9.4	-2.0	-3.4	-2.3	-7.5
9	-6.3	-5.9	-8.4	-5.8	-3.6
10	-4.5	-3.3	-7.5	-6.0	-7.5
11	-8.7	-6.3	-9.1	-1.9	-7.4
12	-7.0	-5.5	-2.2	-1.9	-1.5
13	-8.2	-5.8	-2.9	-2.4	-4.4
14	-9.4	-1.9	-5.6	-9.3	-9.8
15	-6.1	-6.6	-2.3	-4.7	-7.7
16	-0.9	-7.0	-0.8	-8.2	-7.5
17	-8.9	-4.1	-9.6	-0.7	-1.3
18	-4.4	-1.0	-0.3	-9.7	-3.0
19	-1.4	-7.6	-7.7	-6.7	-3.3
20	-0.3	-3.3	-7.3	-5.7	-7.5

b:

b_j	
1	91
2	110
3	108
4	96
5	105
6	103
7	99
8	97
9	100
10	93
11	101
12	92
13	97
14	108
15	95

A:

a_{jk}	1	2	3	4	5	6	7	8	9	10	11	12	13	14	15	16	17	18	19	20
1	2	8	9	2	0	7	6	1	8	4	8	3	7	3	4	1	4	4	3	0
2	3	5	2	4	8	8	8	2	3	1	2	8	7	9	8	0	9	8	5	9
3	2	3	0	3	1	8	5	6	0	1	7	9	3	5	6	9	0	5	0	0
4	9	5	6	2	8	7	9	7	9	5	2	7	8	2	2	3	7	5	8	2
5	8	8	2	6	2	1	0	9	5	6	9	8	1	4	9	3	1	2	8	8
6	0	7	5	8	6	9	6	2	7	7	4	9	9	2	0	1	3	8	9	8
7	4	0	9	4	6	1	2	5	4	8	7	7	8	6	5	9	9	4	1	8
8	2	6	4	2	8	8	7	8	1	0	8	4	1	6	2	2	4	7	0	8
9	5	8	3	7	0	8	6	9	7	4	5	9	3	6	2	0	2	0	3	0
10	2	9	7	7	1	7	2	1	0	1	8	3	1	1	3	6	2	2	6	2
11	2	0	2	3	4	8	2	1	5	8	8	0	2	1	0	1	3	7	2	3
12	7	7	3	9	9	1	7	2	5	1	9	7	0	7	2	3	4	5	1	2
13	2	9	0	2	3	8	4	2	5	7	4	0	1	5	0	0	4	1	5	0
14	1	5	1	4	7	3	7	8	7	8	7	7	3	2	0	8	2	3	5	9
15	4	2	7	5	5	1	6	1	7	9	3	9	3	2	0	5	2	6	2	7

Test 2

$\psi_k = 5$ for $k=1,\dots,20$

F:

f_{ks}	1	2	3	4	5
1-20	like in Test 1				

b:

b_j	
1	105
2	118
3	123
4	138
5	110
6	127
7	111
8	108
9	121
10	133
11	134
12	138
13	122
14	111
15	132
16	107
17	130
18	100
19	139
20	104

A:

a_{jk}	1	2	3	4	5	6	7	8	9	10	11	12	13	14	15	16	17	18	19	20
1	1	6	1	9	4	1	4	9	6	9	9	6	0	6	1	9	4	0	8	0
2	9	0	4	7	2	8	0	1	2	9	1	9	9	8	7	2	6	3	0	3
3	6	5	0	1	2	5	0	2	0	7	4	1	0	2	4	5	8	4	1	7
4	8	2	0	7	4	7	1	6	1	6	7	5	3	0	7	0	3	8	1	4
5	6	1	9	4	5	9	1	7	9	2	9	3	7	4	6	9	9	1	4	9
6	2	4	3	9	7	5	0	0	6	7	5	2	7	2	7	1	3	3	8	3
7	1	8	0	0	8	5	4	6	1	6	1	9	8	8	7	2	1	9	0	5
8	2	6	9	0	8	1	2	4	2	9	4	5	9	6	9	1	8	2	2	4
9	7	6	4	6	6	6	9	5	4	9	1	1	6	5	8	0	3	0	0	5
10	3	0	1	5	3	0	6	8	4	7	4	5	5	9	7	7	9	4	2	8
11	1	2	9	1	3	3	5	1	9	5	6	4	6	6	5	2	7	0	7	0
12	1	1	8	2	9	2	1	4	5	3	3	8	9	3	3	7	6	8	8	8
13	6	3	6	7	2	3	0	8	3	3	8	2	9	9	5	1	9	3	7	1
14	6	1	6	8	6	6	4	4	4	8	1	9	0	6	6	9	1	1	6	8
15	4	5	6	3	8	9	4	7	0	3	3	0	7	0	0	1	7	4	2	1
16	2	1	8	6	6	7	3	2	1	8	7	8	0	2	9	3	0	7	0	3
17	6	9	9	2	2	1	2	0	3	5	5	3	8	0	5	0	4	6	3	8
18	6	1	5	0	3	5	9	4	3	1	8	5	7	4	4	1	5	5	3	6
19	3	9	2	0	9	7	0	9	6	2	6	5	1	7	5	7	1	5	3	4
20	5	9	0	8	4	9	5	1	0	6	8	4	8	0	6	9	6	3	6	3

Test 3

$$\psi_k = 5 \quad \text{for} \quad k=1,\dots,5$$

F:

f_{ks}	1	2	3	4	5
1	-8.8	-6.1	-5.6	-7.5	-9.9
2	-8.6	-8.0	-3.3	-7.3	-5.0
3	-5.1	-2.9	-6.7	-7.8	-3.7
4	-7.7	-5.9	-3.4	-6.5	-7.7
5	-4.7	-3.6	-5.2	-7.5	-4.7

A:

a_{jk}	1	2	3	4	5
1	5	9	0	0	1
2	4	2	3	1	3
3	2	3	5	1	9

b:

b_j	
1	19
2	28
3	35

Test 4

$$\psi_k = 5 \quad \text{for} \quad k=1,\dots,10$$

F:

f_{ks}	1	2	3	4	5
1-5	like in Test 3				
6	-6.3	-4.5	-4.0	-3.5	-1.9
7	-2.3	-6.8	-7.2	-9.7	-7.1
8	-9.4	-5.2	-8.9	-9.5	-1.3
9	-2.8	-0.1	-4.5	-0.5	-7.9
10	-5.7	-9.9	-9.0	-3.3	-8.4

A:

a_{jk}	1	2	3	4	5	6	7	8	9	10
1	7	7	5	2	0	1	3	7	9	4
2	1	6	0	2	8	0	4	1	9	6
3	9	7	6	1	9	5	4	7	0	3
4	2	6	5	5	8	0	8	8	1	8
5	4	9	1	6	1	7	4	2	0	8

b:

b_j	
1	117
2	102
3	125
4	153
5	114

Lecture Notes in Control and Information Sciences

Edited by M. Thoma

Vol. 43: Stochastic Differential Systems
Proceedings of the 2nd Bad Honnef Conference
of the SFB 72 of the DFG at the University of Bonn
June 28 – July 2, 1982
Edited by M. Kohlmann and N. Christopeit
XII, 377 pages. 1982.

Vol. 44: Analysis and Optimization of Systems
Proceedings of the Fifth International
Conference on Analysis and Optimization of Systems
Versailles. December 14–17, 1982
Edited by A. Bensoussan and J. L. Lions
XV, 987 pages, 1982

Vol. 45: M. Arató
Linear Stochastic Systems
with Constant Coefficients
A Statistical Approach
IX, 309 pages. 1982

Vol. 46: Time-Scale Modeling of Dynamic Networks
with Applications to Power Systems
Edited by J. H. Chow
X, 218 pages. 1982

Vol. 47: P. A. Ioannou, P. V. Kokotovic
Adaptive Systems with Reduced Models
V, 162 pages. 1983

Vol. 48: Yaakov Yavin
Feedback Strategies for Partially
Observable Stochastic Systems
VI, 233 pages, 1983

Vol. 49: Theory and Application of Random Fields
Proceedings of the IFIP-WG 7/1
Working Conference
held under the joint auspices of the
Indian Statistical Institute
Bangalore, India, January 1982
Edited by G. Kallianpur
VI. 290 pages. 1983

Vol. 50: M. Papageorgiou
Applications of Automatic Control Concepts
to Traffic Flow Modeling and Control
IX, 186 pages. 1983

Vol. 51: Z. Nahorski, H.F. Ravn, R.V.V. Vidal
Optimization of Discrete Time Systems
The Upper Boundary Approach
V, 137 pages 1983

Vol. 52: A. L. Dontchev
Perturbations, Approximations and Sensitivity Analysis
of Optimal Control Systems
IV, 158 pages. 1983

Vol. 53: Liu Chen Hui
General Decoupling Theory of Multivariable
Process Control Systems
XI, 474 pages. 1983

Vol. 54: Control Theory for Distributed
Parameter Systems and Applications
Edited by F. Kappel, K. Kunisch,
W. Schappacher
VII, 245 pages. 1983.

Vol. 55: Ganti Prasada Rao
Piecewise Constant Orthogonal Functions
and Their Application to Systems and Control
VII, 254 pages. 1983.

Vol. 56: Dines Chandra Saha, Ganti Prasada Rao
Identification of Continuous
Dynamical Systems
The Poisson Moment Functional
(PMF) Approach
IX, 158 pages. 1983.

Vol. 57: T. Söderström, P. G. Stoica
Instrumental Variable Methods
for System Identification
VII, 243 pages. 1983.

Vol. 58: Mathematical Theory of
Networks and Systems
Proceedings of the MTNS-83 International
Symposium
Beer Sheva, Israel, June 20–24, 1983
Edited by P. A. Fuhrmann
X, 906 pages. 1984

Vol. 59: System Modelling and Optimization
Proceedings of the 11th IFIP Conference
Copenhagen, Denmark, July 25-29, 1983
Edited by P. Thoft-Christensen
IX, 892 pages. 1984

Vol. 60: Modelling and Performance
Evaluation Methodology
Proceedings of the International Seminar
Paris, France, January 24–26, 1983
Edited by F. Bacelli and G. Fayolle
VII, 655 pages. 1984

Vol. 61: Filtering and Control of Random
Processes
Proceedings of the E.N.S.T.-C.N.E.T. Colloquium
Paris, France, February 23–24, 1983
Edited by H. Korezlioglu, G. Mazziotto, and
J. Szpirglas
V, 325 pages. 1984

Lecture Notes in Control and Information Sciences

Edited by M. Thoma and A. Wyner

Vol. 62: Analysis and Optimization
of Systems
Proceedings of the Sixth International
Conference on Analysis and Optimization
of Systems
Nice, June 19–22, 1984
Edited by A. Bensoussan, J. L. Lions
XIX, 591 pages. 1984.

Vol. 63: Analysis and Optimization
of Systems
Proceedings of the Sixth International
Conference on Analysis and Optimization
of Systems
Nice, June 19–22, 1984
Edited by A. Bensoussan, J. L. Lions
XIX, 700 pages. 1984.

Vol. 64: Arunabha Bagchi
Stackelberg Differential Games
in Economic Models
VIII, 203 pages, 1984

Vol. 65: Yaakov Yavin
Numerical Studies
in Nonlinear Filtering
VIII, 273 pages, 1985.

Vol. 66: Systems and Optimization
Proceedings of the Twente Workshop
Enschede, The Netherlands, April 16–18, 1984
Edited by A. Bagchi, H. Th. Jongen
X, 206 pages, 1985.

Vol. 67: Real Time Control of Large Scale Systems
Proceedings of the First European Workshop
University of Patras, Greece, Juli 9–12, 1984
Edited by G. Schmidt, M. Singh, A. Titli,
S. Tzafestas
XI, 650 pages, 1985.

Vol. 68: T. Kaczorek
Two-Dimensional Linear Systems
IX, 397 pages, 1985.

Vol. 69: Stochastic Differential Systems –
Filtering and Control
Proceedings of the IFIP-WG 7/1 Working Conference
Marseille-Luminy, France, March 12-17, 1984
Edited by M. Metivier, E. Pardoux
X, 310 pages, 1985.

Vol. 70: Uncertainty and Control
Proceedings of a DFVLR International Colloquium
Bonn, Germany, March, 1985
Edited by J. Ackermann
IV, 236 pages, 1985.

Vol. 71: N. Baba
New Topics in Learning Automata
Theory and Applications
VII, 231 pages, 1985.

Vol. 72: A. Isidori
Nonlinear Control Systems:
An Introduction
VI, 297 pages, 1985.

Vol. 73: J. Zarzycki
Nonlinear Prediction
Ladder-Filters for Higher-Order
Stochastic Sequences
V, 132 pages, 1985.

Vol. 74: K. Ichikawa
Control System Design based on
Exact Model Matching Techniques
VII, 129 pages, 1985.

Vol. 75: Distributed Parameter
Systems
Proceedings of the 2nd International
Conference, Vorau, Austria 1984
Edited by F. Kappel, K. Kunisch,
W. Schappacher
VIII, 460 pages, 1985.

Vol. 76: Stochastic Programming
Edited by F. Archetti, G. Di Pillo,
M. Lucertini
V, 285 pages, 1986.

Vol. 77: Detection of
Abrupt Changes in Signals
and Dynamical Systems
Edited by M. Basseville,
A. Benveniste
X, 373 pages, 1986.

Vol. 78: Stochastic
Differential Systems
Proceedings of the 3rd Bad Honnef
Conference, June 3–7, 1985
Edited by N. Christopeit, K. Helmes,
M. Kohlmann
V, 372 pages, 1986.

Vol. 79: Signal
Processing for Control
Edited by K. Godfrey, P. Jones
XVIII, 413 pages, 1986.

Vol. 80: Artificial Intelligence
and Man-Machine Systems
Edited by H. Winter
IV, 211 pages, 1986.

Lecture Notes in Control and Information Sciences

Edited by M. Thoma and A. Wyner

Vol. 81: Stochastic Optimization
Proceedings of the International Conference,
Kiev, 1984
Edited by V. I. Arkin, A. Shiraev, R. Wets
X, 754 pages, 1986.

Vol. 82: Analysis and Algorithms
of Optimization Problems
Edited by K. Malanowski, K. Mizukami
VIII, 236 pages, 1986.